Communicating the Climate Crisis

Environmental Communication and Nature: Conflict and Ecoculture in the Anthropocene

Series Editor: C. Vail Fletcher, University of Portland

This interdisciplinary book series seeks original proposals that examine environmental communication scholarship. In the Anthropocene era, the period during which human activity has become the dominant influence on climate and the environment, the need for highlighting and re-centering nature in our worldviews and policies is urgent, as collapsing ecosystems across the globe struggle to survive. Topics might include climate change, land use conflict, water rights, natural disasters, non-human animals, the culture of nature, ecotourism, wildlife management, human/nature relationships, food studies, sustainability, eco-pedagogy, mediated nature, eco-terrorism, environmental education, ecofeminism, international development, and environmental conflict. Ultimately, scholarship that addresses the general overarching question "how do individuals and societies make sense of and act against/within/out of nature?" is welcomed. This series is open to contributions from authors in environmental communication, environmental studies, media studies, rhetoric, political science, critical geography, critical/cultural studies, and other related fields. We also seek diverse and creative epistemological and methodological framings that might include ethnography, content analysis, narrative and/or rhetorical analysis, participant observation, and community-based participatory research, among others. Successful proposals will be accessible to a multidisciplinary audience.

Recent Titles in This Series

Water, Rhetoric, and Social Justice: A Critical Confluence Edited by Casey R. Schmitt, Theresa R. Castor, and Christopher S. Thomas

Environmental Activism, Social Media, and Protest in China: Becoming Activists over Wild Public Networks By Elizabeth Brunner

Natural Disasters and Risk Communication: Implications of the Cascadia Subduction Zone Megaquake Edited by C. Vail Fletcher and Jennette Lovejoy

Critical Environmental Communication: How Does Critique Respond to the Urgency of Environmental Crisis? By Murdoch Stephens

Natural Disasters and Risk Communication: Implications of the Cascadia Subduction Zone Megaquake Edited by C. Vail Fletcher and Jennette Lovejoy

Communicating in the Anthropocene: Intimate Relations Edited by C. Vail Fletcher and Alexa Dare

Communicating the Climate Crisis: New Directions for Facing What Lies Ahead By Julia B. Corbett

Communicating the Climate Crisis

New Directions for Facing What Lies Ahead

Julia B. Corbett

LEXINGTON BOOKS
Lanham • Boulder • New York • London

Published by Lexington Books
An imprint of The Rowman & Littlefield Publishing Group, Inc.
4501 Forbes Boulevard, Suite 200, Lanham, Maryland 20706
www.rowman.com

6 Tinworth Street, London SE11 5AL, United Kingdom

Copyright © 2021 The Rowman & Littlefield Publishing Group, Inc.

Cover Image: UN photo by Logan Abassi. A father carries his daughter on his shoulders as residents flee rising waters after heavy rains caused by tropical storm "Noel" flooded their homes in Cité Soleil, Haiti, in 2007. Used with permission.

All rights reserved. No part of this book may be reproduced in any form or by any electronic or mechanical means, including information storage and retrieval systems, without written permission from the publisher, except by a reviewer who may quote passages in a review.

British Library Cataloguing in Publication Information Available

The hardback edition of this book was previously catalogued by the Library of Congress as follows:

Library of Congress Cataloging-in-Publication Data

Names: Corbett, Julia B., author.
Title: Communicating the climate crisis : new directions for facing what lies ahead / Julia B. Corbett.
Description: Lanham : Lexington, [2021] | Series: Environmental communication and nature : conflict and ecoculture in the anthropocene | Includes bibliographical references and index. | Summary: "Communicating the Climate Crisis lays out fresh directions and strategies for creating a new story of hope through action-not as isolated and "guilty" consumers, but as social actors who use emotional resilience, climate conversations, justice, and faith to break the current social inertia and create a desired future" —Provided by publisher.
Identifiers: LCCN 2020054357 (print) | LCCN 2020054358 (ebook) | ISBN 9781793638021 (cloth) | ISBN 9781793638045 (paperback) | ISBN 9781793638038 (epub)
Subjects: LCSH: Climatic changes. | Communication in climatology. | Climatic changes—Public opinion. | Climate change mitigation.
Classification: LCC QC902.9 .C67 2021 (print) | LCC QC902.9 (ebook) | DDC 363.738/7452—dc23
LC record available at https://lccn.loc.gov/2020054357
LC ebook record available at https://lccn.loc.gov/2020054358

*To the countless individuals worldwide who work compassionately and tirelessly to secure protection and love of this good Eairth.
You are my inspiration and my hope.*

Contents

Acknowledgments	ix
1 E*air*th	1
2 Fossil Fuel Culture	25
3 Individuals as Social Actors, Not Consumers	53
4 Emotions and Climate Silence	79
5 Breaking the Silence: Strategies for Talking about Climate Change	109
6 Justice and Faith: The Moral Imperative of Climate Change	137
7 A New Relationship with E*air*th	167
8 Telling a New Story	191
Index	213
About the Author	219

Acknowledgments

First and foremost, I acknowledge the scores of scholars whose dedicated work broadened my view of the climate crisis and inspired new paths forward. It was rewarding to wander in and out of my communication home and learn of our vast connections across disciplines. Your names and contributions fill the vast endnotes.

The guest speakers in Fall 2013 greatly helped me develop the climate communication class that grew into this book: Robert Brulle, Sharon Dunwoody, Ed Maibach, Susanne Moser, and locally Andrea Brunelle, Jim Steenburgh, Simon Brewer, Judy Fahys, Matthew LePlante, Terry Gildea, and Glen Feighery. Thanks to the Council of Dee Fellows for the Dee Grant, which brought the speakers to campus and generated widespread attendance at these events.

On campus, thanks to my colleagues at the Global Change and Sustainability Center and the Environmental Humanities Graduate Program for their work and conviviality (in particular, Brett Clark for his steadfast support and friendship and Lizzie Calloway for the introduction to Zotero!). The EH graduate students in my spring 2018 seminar graciously read a few draft chapters and reams of readings. Although Hollywood never optioned our pilot for *Lignite*, it was one of the most enjoyable, creative, and exceptional group projects of my career. Your passion, drive, and heart – and that of all EH students – are a continuing source of hope. I'm grateful to the University of Utah for the Sabbatical to help complete this manuscript. Thanks to my undergrads who helped test-run all eight chapters in spring 2020; I appreciate your dedication and vulnerability to engage this difficult topic and endure the eventual transition to online.

Thanks for permission to print the Global Warming's Six Americas graphic; the respective climate change communication centers at George

Mason and Yale universities produce such important and valuable work. Thanks to Laura Schmidt (former EH student!) for permission to print the ten steps of grief work by Good Grief; the emotions of the climate crisis touch us all.

A big hug to friends near and far for your love and support; you understood that this was a hard but important book to write. Debora Threedy, you were wonderful to critically read several chapters. And sister Sara, your love has so buoyed me during the perfect storm of events over the last few years.

And to the good Eairth; you remind me that the reason I'm here is to nurture love of the living world.

Chapter 1

E*air*th

If I asked you, "Where do you live?" you would probably name your city or your street address – a physical place on the ground. This is how we tend to think about our planet: it's all about life on the surface – the places we plant our feet and flowers, build our houses, and traverse for work and leisure (which might include watery surfaces of lakes and oceans). Life seems to be about living *on* the "crust."

But here's a different perspective about where you live. Go outside, lie down, and gaze up. If there are clouds up there, watch them build, billow, drift. Right now, you are spinning through space at about 1,000 miles per hour (depending on your latitude), but it doesn't feel very fast because gravity holds you firmly to the ground. How fast do you suppose the clouds are moving? A couple of miles an hour, maybe twenty miles an hour? The point is: they are *not* hurtling by at a thousand miles an hour because they are as much a *part* of the Earth as you are.

Well, of course, you know that: life on Earth includes the air you breathe and those clouds. But nevertheless, we don't identify with "up there" the same way we do with life "down here." We don't tend to think of ourselves as living *within* an atmosphere.

Philosopher David Abram suggested we add an *i* to our planet's name – *Eairth* (and yes, it sounds Scottish when he says it) – to remind us that we live *in* this "amniotic substance" and not merely *on* the surface:

> The air is not a random bunch of gases simply drawn to earth by the earth's gravity, but an elixir generated by the soils, the oceans, and the numberless organisms that inhabit this world, each creature exchanging certain ingredients for others as it inhales and exhales, drinking the sunlight with our leaves or filtering the water with our gills, all of us contributing to the composition of this

phantasmagoric brew.... It is as endemic to the earth as the sandstone beneath my boots.[1]

What we call "air" is a product made and moved by all of us – ants, humans, trees, bacteria, plankton, zebras – all interacting within the amniotic atmosphere.

Now, look skyward again for a distance of about six miles or 33,000 feet (to help, imagine someplace that's six miles away, like a friend's house or your workplace). Even the highest clouds you see are within that six-mile span. A jet contrail is sometimes a little bit higher (if the pilot wanted to avoid rough air). Even with clouds or contrails, it's really hard to judge the distance because the sky seems empty and without depth. Abram calls *depth* the dimension of closeness and distance, a "visceral stretch between the near and far of things." In an Ansel Adams Yosemite photograph, you see the distant mountain as enormous because of the space and depth the "empty" air provides between photographer and mountain. Even as children, we grasp our spatial relationship with air in crayon drawings of blue sky with puffy clouds and a rayed sun high above the land.

Everyone feels attached to certain places and features on the crust; in contrast, air can feel anonymous and impersonal. So, while gazing skyward, think of the air surrounding your body as your *personal* air space, close and accessible. It fills your lungs and blood and speech; you absorb moisture from it; you move your arm and foot through it. You experience air in daily life in wondrous ways: a windy muss of hair, rustling leaves, cloud sculptures, colors of dawn, enveloping fog, and birds riding warm updrafts as they spiral higher and higher. Although some say, "We are what we eat," in a more essential way, *we are what we breathe* and in what we are continually immersed—an atmosphere. The point of this sky-gazing exercise: all of the elements of life you experience on the surface—the grass or bench upon which you lie, your clothes, the food in your stomach—are possible because of what happens in the six miles above you.

The first time any of us saw an "outsider" view of the entire Eairth was in the "Blue Marble" photo taken by an Apollo 17 astronaut in 1972.[2] In this magnificent photo, our planet's surface looked rather fragile and vulnerable. What is not apparent is what protects the Earth: its atmosphere, the largely invisible layer of gases and moisture enveloping the planet that makes life possible. The globe in every classroom also shows a planet without its atmosphere, which would be as thin as a sheet of paper enveloping it. It's in the atmosphere where the dynamics of climate changes take place.

Next is a discussion about the science of climate change (sometimes neglected in climate communication books) and what it means for life on the crust, now and in the future. This is the only chapter with much science, which is a crucial backdrop for communicating about climate change. It explains the evidence for how and why climate change is happening right now, the documented consequences, and the scientific agreement about it. Finally, it outlines the book's chapters about communication challenges and new directions.

THE SCIENCE OF CLIMATE CHANGE

Several years ago, I joined thousands of people across the US to watch the predicted total solar eclipse. Every day, we take numerous scientific predictions or findings (like the eclipse) to the bank. You also trust that the sun will "rise" tomorrow morning.[3] When you throw a ball in the air, thanks to gravity, you know it will fall back down. You accept that viruses cause colds (or worse). That a snowstorm is headed your direction. That too much solar radiation can cause skin cancer. That DNA carries hereditary information.

But one arena of scientific findings (for a variety of reasons that we will discuss) has not been accepted by a very small minority: human actions have brought about *climate change*. There is now more scientific research related to the phenomenon of climate change than for most topics ever studied (except for cancer), and, scientists are extremely certain (97–98 percent) that it's happening and that humans are responsible. A good test of any scientific theory is whether what's predicted ultimately occurs, and for climate change many of the predictions have already "come true" – and often at a much faster pace and to a greater extent than originally predicted.[4]

In your sky-gazing, you looked up through the 6.2 miles of the lower atmospheric zone called the *troposphere*. This is where weather happens and moisture cycles. The movement of air and moisture in the troposphere is aided by the fact that Eairth tilts toward (in summer) or away from (in winter) the sun as it spins through its twenty-hour-cycle, heating some areas and cooling others, and spawning photosynthesis. Warmed air rises (that's why the second story of a house is warmer than the first) and it carries more moisture. What's flowing and moving above you is intricately connected to—and responsible for—*everything* rooted down here. In the troposphere, various gases act as a gatekeeper of sorts for how much solar energy (aka sunshine, which we feel as heat) reaches the surface. Eairth is a Goldilocks planet: not too hot, not too cold. The troposphere's gases allow some heat to escape back into space at night so we don't get too hot (like boiling Venus), but the gases trap enough heat near the surface so we don't get too cold (like freezing Mars). The delicate exchange of heat energy between the troposphere and Earth's surface means you get a livable planet with predictable temperature ranges. Thus, thanks for anything loved "down here"—family, water, mountains, ice cream, pets—should go to the atmosphere swirling about your head.

So how did the heat-regulator function of the troposphere get messed up? When humans burn great amounts of fossil fuels (coal and petroleum products such as oil, natural gas, and gasoline), it releases a lot of *greenhouse gases* (primarily carbon dioxide, but also other gases) into the troposphere. Small amounts of these gases in the troposphere are crucial for balancing the sun's incoming and outgoing heat energy, but when excessive amounts get trapped up there, they act like a blanket around the Earth, which warms it up. Some people call this the *greenhouse effect* because it's similar to how the glass roof on a greenhouse lets sunshine in and then traps the heat inside, which warms the plants. For Eairth, adding too many greenhouse gases to the troposphere means the greenhouse effect is stronger and heat energy has trouble escaping back to space.

Warmer global temperatures are the initial result of that hot "greenhouse," which is why some people use the term *global warming* or *global heating*. Each month, you are likely to hear of a new record high temperature, and sometimes many records (2010–2019 was the hottest decade on record and all of its years broke records). But in addition to warmer temperatures, the "greenhouse" heating disrupts how the climate system operates, which is why *climate change* is a more comprehensive term for all the ways that warming affects weather and water cycling. Increasingly, journalists and others use "climate crisis" or "climate emergency" to stress the urgency and severity of changes to the global climate system.

Scientists—as well as lay folks like us—have already seen across the globe the predicted effects of climate change when compared to previous decades and centuries, such as:

- increasing warm temperatures (record highs and record warm lows),
- changing precipitation patterns (such as more intense and/or frequent droughts, more frequent and heavy rainfall and floods, and less snow that also melts sooner),
- rising sea levels (half from thermal expansion and half from melting land-based ice), now rising twice as fast as in the 1990s, and flooding island nations and dozens of coastal cities at high tides,
- more extreme weather events (such as hurricanes that stall and dump more rain), and
- changes in ocean mixing, currents, and declining oxygen content and increasing acidity (already one-third more acidic).

Other climate-linked impacts we already experience include a growth in acres and severity of wildfires (from Australia to California), worsening air pollution (both from increased temperatures and wildfire smoke), spread of infectious diseases (aided by warming temperatures), decreased agricultural productivity, and growing food and water scarcity.

Another element important to communicating climate change is the distinction between "weather" and "climate." *Weather* is the atmospheric conditions for an individual moment—say, what you encounter when you walk out the front door (hot and dry, or a rain shower). In the day-to-day weather you experience, a couple of degrees warmer one day and cooler the next is no big deal—in fact it's typical. But *climate* is the average of weather conditions over a very long period of time—like several decades or centuries. When the TV meteorologist says today's high temperature and precipitation were far "above normal," the normals were established through *climate* records. As some people put it: climate is what you expect; weather is what you get. And of course, my regional climate in the US Intermountain West is different from the climate in San Francisco or San Paulo, each affected by factors such as the amount of sunlight or clouds, elevation, latitude, topography, and distance from oceans. Global climate change is all the disruptions resulting from more heat being trapped, averaged from climates all over the world and over a very long period of time.

The Consilience of Evidence

Weather fluctuates every day, while climate—not so much, particularly since humans have lived on Eairth. Sure, over the past several centuries regions have experienced long droughts or decades that were colder or hotter. But when scientists say that Eairth's climate is changing, they mean compared to the *very* long term, like many thousands of years. In that sense, the climate is currently changing faster than any time in the history of human civilization on this planet. Evidence for this comes from different, independent data sources worldwide that reach the same conclusion (what is called a *consilience of evidence*): global temperatures are rising and the climate is changing as a result of human actions.[5] Here is a look at some of that evidence.

When you want to know the temperature, you consult a thermometer. But how can scientists track meteorological conditions before instruments like thermometers were invented? They switch from instrumental records to the Eairth's natural record systems of past climate conditions, trapped and recorded in sediments (such as in lakes), ice cores from glaciers and ice sheets, tree rings, coral, and temperatures deep in the earth (collected as bore-hole data).

The world's permanent ice are essentially records of very "old air" trapped in the ice, similar to tiny air bubbles you see trapped in an ice cube or frozen lake. By drilling and removing long cylinders of ice that have been compressed for thousands of years (such as in a glacier), scientists can measure the concentration of gases in the "old air," including the most abundant greenhouse gas, carbon dioxide. The ice cores reveal that the concentration of CO_2 in the air oscillated between 180 and 280 parts per million (ppm)—from about 800,000 years ago up until about 300 years ago. That consistency is truly remarkable given the ways that climate has changed during these eons (such as the last ice age that ended about 12,000 years ago) and still the lower atmosphere balanced the greenhouse gases to these concentrations.

So, what happened about 300 years ago? The Industrial Revolution commenced and humans began burning enormous amounts of wood and coal (and later, oil and natural gas). This era also brought great deforestation; forests (like oceans) are vital storehouses or "sinks" of carbon. Also during this time, the world's population began bumper growth, from about 1 billion in 1800 to nearly 8 billion today.

Today, human activities release a whopping 40 billion tons of carbon dioxide into the atmosphere each year, and the planet's oceans and lands can absorb only a fraction of that. In March 2020, the concentration of CO_2 in the troposphere topped 415 ppm—far above the 280-ppm high-mark that held for over 800,000 years[6] and a concentration that last occurred about 15 million years ago.[7] The thicker blanket has indeed increased temperatures. The global average temperature (on land and ocean) has warmed nearly 2 degrees Fahrenheit since 1901 (with two-thirds of that warming since 1975). However, warming does not happen equally across the planet; temperature increases across Alaska and the Arctic are more than double the global average (4 degrees Fahrenheit, and higher), contributing to the lowest extent of sea ice ever recorded and the thawing of the "permafrost." On February 8, 2020, icy Antarctica had the warmest temperature ever recorded: 65 degrees Fahrenheit; that June 20, a Siberian Arctic town hit a record of 100 degrees Fahrenheit.[8] In the mainland US, the region with the greatest temperature increases is the Southwest (the fastest warming city is Las Vegas, which has warmed 5.8 degrees Fahrenheit since 1970).

If there's one chart to remember about climate change, this is it (figure 1.1). You'll notice that the lines are jagged, not smooth, because, of course, both weather and climate do indeed fluctuate, season-to-season and year-to-year. The downward lines are what climate deniers have focused on: when temperatures cooled. This is called "cherry-picking" the data because it makes the results seem to reach a different conclusion when they are picked from the bigger picture. After 400,000 years, the steep, rapid increase of CO_2 from the 1950s onward is undeniable: concentrations of greenhouse gases like carbon dioxide have altered the ability of the troposphere to balance incoming and outgoing heat energy, disrupting climate cycles, and rapidly warming the Eairth. No matter how you look at it, this is in no way a "normal" fluctuation.

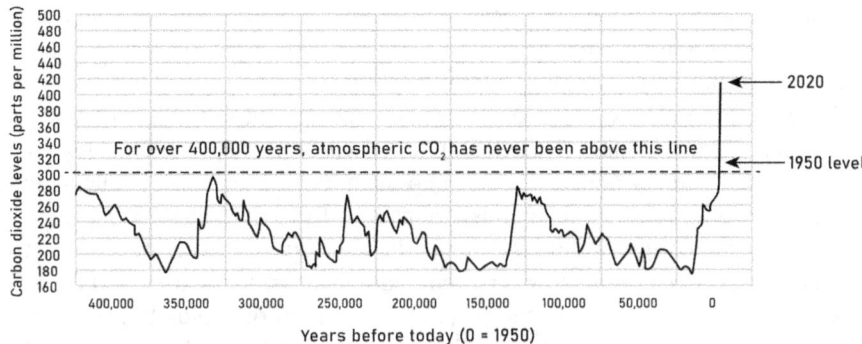

Figure 1.1 This Graph, Based on the Comparison of Atmospheric Samples Contained in Ice Cores and More Recent Direct Measurements, Provides Evidence That Atmospheric CO_2 Has Increased Dramatically Since the Industrial Revolution. Source: NASA, https://climate.nasa.gov/evidence/, Luthi, D., et al. 2008; Etheridge, D.M., et al. 2010; Vostok ice core data/J.R. Petit et al.; NOAA Mauna Loa CO2 Record. Graph Update by Mark Draper.)

IT'S US

Sometimes I hear people say that there is no way that humans could alter the climate of the entire planet; I agree that it seems preposterous, but it all depends on how the troposphere functions. Even "small" changes in the concentration of greenhouse gases up there can profoundly alter how thick the blanket is and how much heat it traps. Like in a finely tuned machine, it doesn't take much to gum up the engine. And, humans have a long track record of our ability to profoundly alter the Earth's surface. In the 50,000 years that "modern" humans have walked the planet, virtually no region or ecosystem has been left unaltered, whether from land clearing and deforestation, development, mining, and in just the last two centuries, an exploding population. One estimate is that humans have altered over 70 percent of the planet's ice-free surface,[9] so it is not surprising that humans can just as profoundly alter Eairth as they have Earth.

And, there is no scientific evidence for alternative explanations as to why the climate has changed—particularly, so incredibly quickly. Fluctuations in solar output and Eairth's natural variability (such as El Niño events) do not contribute to well-documented net warming climate trends. A multivolume report by fourteen scientific agencies concluded that human influence has been by far the dominant cause of warming in the last century; our current climate changes are decidedly *anthropogenic*, or human-caused.[10]

I understand why it's tempting to think that humans are not causing climate change. If it were somehow "natural," then we're not responsible and there's nothing we need to do and no changes we need to make. Yet, the only valid explanation is that humans' prodigious combustion of fossil fuels is causing global warming and disrupting climatic cycles.

So, where do all these greenhouses gases come from? Let's start with an individual's carbon output. Each gallon of gasoline you burn in your car releases about twenty pounds of CO_2 into the atmosphere. (I know that sounds impossible; gas weighs 6.3 pounds per gallon. But when burned, the carbon bonds with two oxygen atoms, which nearly quadruples its weight. For a robust explanation, check out this link.[11]) Then imagine the gasoline used and the CO_2 released by all the drivers in the planet's population of 7-plus billion. That's weighty. And, that doesn't include all the energy used to make and deliver all the products you use at home and at work, nor the energy used to heat and cool those places.

By one account, the average American emits 18 million tons of CO_2 each year (while the average human on the planet emits just 4 million tons).[12] One way to visualize your personal contribution: imagine walking around each day surrounded by a brand *new*, 12-foot, 110-pound sphere of your CO_2 emissions. That's even weightier. So yes, our actions are that "powerful" (particularly the seventy corporations responsible for two-thirds of carbon emissions since the industrial era). Some scientists now call humans "geological agents," able to change the physical processes of the Earth,[13] including the operation of the climate system.

Scientists Agree

Another thing I hear some people say is that scientists don't agree about climate change and there's too much uncertainty in the science. That message, brought to you by industry-funded campaigns (and politicians), is specifically crafted to raise doubts and uncertainty in the public's mind so that no action is taken on this urgent crisis. Virtually all of the world's credentialed scientists (97 percent) accept the reality of human-induced climate change, and no bona fide scientists have replicable published studies (in professional science journals) that refute that climate change is occurring and that human activities produce the vast majority of the emissions that cause it.[14] (How might you evaluate a 97 percent agreement in other areas of your life, whether a disease diagnosis or auto safety records?)

So why do you keep hearing about disagreement? Much of that has to do with where you read and hear about climate change. Climate denial is far greater in the United States and Australia than in most other countries. The handful of US climate deniers have sought (and received) a great deal of attention in mass media (and social media), actively publishing opinion pieces and serving as a source for reporters who feel the need to publicize "both sides," which creates what some scholars call an "ersatz balance"[15] and distorts the 97 percent scientific agreement. Most of all deniers are not climate scientists nor experts on what they critique, but they are extremely active in writing (non-peer-reviewed) books, blogs, and videos. Although some oil and gas companies knew in the early 1980s of the risks of catastrophic climate change, they did not reveal this to the public nor to shareholders and instead spent millions to spread false messages of scientific "uncertainty." ExxonMobil, for example, faces investigations by multiple US state attorneys general about how they falsely communicated the risk of climate change to investors and the public. (See this note for readings on the "climate denial" movement and where to obtain reliable climate science.[16]) Another important piece of the "uncertainty" puzzle is that psychologically we don't want to accept the reality of climate change, which a later chapter delves into.

As scientist and author Naomi Oreskes said, "Is there disagreement over the details of climate change? Yes. Are all aspects of climate past and present well understood? No, but whoever claimed that they were? Does climate science tell us what policy to pursue? Definitely not, but it does identify the problem, explain why it matters, and give society insights that can help to frame an efficacious policy response."[17]

Oreskes identifies two primary areas where the science is not yet certain: *tempo* and *mode* – that is, how fast climate change impacts and changes will occur, and where and how they will manifest. Also under study are the many *feedback* loops within Eairth's complex climate systems. One example is ice and its ability to reflect solar energy. The reflectivity of a material is called its *albedo* and ice has a high one. (Your T-shirt color has a slight albedo effect: you'll be cooler wearing a reflective white T on a hot day than a black one.) Of all of the sunlight that reaches Earth, about 30 percent is reflected back to space, some of that by ice and snow that cover both oceans and lands, particularly at the poles. When ice melts from the ocean surface, the now-darker ocean absorbs far more heat than the reflective ice did, which in turns speeds the melting of more ice. This feedback has occurred at a far faster tempo than scientists predicted.

Another feedback loop is forest fires. Warming temperatures dry out soil and plants, making them more susceptible to fires. Burning trees emit a tremendous amount of CO_2. The extensive British Columbia fires in 2017 produced three times the usual amount of annual greenhouse gas emissions for the entire province.[18] This dramatic increase in emissions accelerates a feedback loop: burning releases carbon, leads to warmer temperatures and long periods without rain, which leads to more fires, which releases more carbon into the atmosphere, which leads to even warmer and drier conditions, and more fires.

Yet another dangerous feedback loop is playing out with recent wildfires in the Arctic and Greenland, where they are extremely rare.[19] Fires produce black soot (also known as "black carbon"), which settles on ice sheets. Darkened ice absorbs more heat than white ice, which melts the ice even faster. Given the amount of permafrost and peat in northern landmasses like Greenland, burning there also releases huge amounts of stored methane (essentially, natural gas), which causes more warming and more fires, which in turn creates more blackened ice and more melting. Skiers take note: studies of snowpack in Colorado and Utah found that dark specks in the snow (wind-blown dirt and carbon) absorbed more heat and lead to more rapid snowmelt.[20]

The science makes it clear: climate change isn't something you choose to "believe" in like your lucky baseball cap. It's a matter of accepting the reams of data gathered and evidence tested, just as you do routinely every day for hundreds of things. In terms of the amount of scientific evidence, not "believing" in climate change is akin to not "believing" in cancer. You have personally witnessed the evidence of climate change in your life and in the larger world.

A CLIMATE-CHANGED WORLD

If you were born after 1976, you have never experienced a year of below-normal temperatures; 1976 was the last time the annual average temperature was cooler than the twentieth-century average.[21] The four hottest years ever recorded were the last four in the 2010 decade, an undeniable heating that will have no doubt been broken by the time you read this, as climate change continues to shift the baseline. If you were born after 1976, your perception of what is environmentally "normal" is based on a warmed world. The crux is that each generation perceives the shifted and warmed baseline as the default, a sort of environmental generational amnesia.[22] This presents a communication challenge when people lack the personal experience of very cold winters and dependable seasons.

Yet no matter your age, climate disruptions will be from this time forward a defining feature of your life. The changes are obvious enough that you can engage in data gathering of your own: ask older friends and relatives what climate changes they have experienced in their lifetimes. My elderly neighbors talk about what winters used to be like in Wyoming, and my aunt has never experienced summers such as the recent ones in the Pacific Northwest, where she's lived most of her life. My friend Kevin found photos he took in the 1970s in Newfoundland with shore ice forming in November and snow remaining on a coastal plain in June; these once-typical cold conditions haven't been seen there in decades.

In addition to remembered experiences, *citizen scientists* observe and collect data around the globe about nature. Perhaps the earliest citizen science project was the Christmas Bird Count, conducted annually since 1900. Many citizen scientists note changes in the seasonal timing of plant and animal life-stages known as *phenology* – timing that has been significantly altered by a warming climate. The USA National Phenology Association has over 15 million data records from citizen scientists, including when trees leaf-out in backyards and when birds migrate.[23]

One finding is that spring now arrives as much as three weeks earlier.[24] Plants are budding, insects emerging, and birds migrating in ways that are increasingly mismatched, or asynchronous. The mismatch between bird arrival date and the onset of spring grew by an average of half a day per year between 2001 and 2012 for forty-eight songbird species.[25] Songbirds migrate primarily spurred by changes in daylight, but the primary signal for plants and trees to bud is warming days. That means birds may arrive too late for an insect hatch they were counting on. The data used for bird arrivals in this study were from Cornell University's eBird project, a collection of 500 million sightings worldwide submitted by bird-watchers (now via cell phones with GPS coordinates) since the early 2000s.[26]

Shifting springs and warming temperatures in a climate-changed world have significant consequences for all species (including humans). If fruit trees flower earlier because it's warming earlier, there's a good chance those blossoms could be killed by frost, which happened recently to cherry trees in Michigan, apples in the Midwest, and peaches in Virginia and Utah. In response to warming, animals tend to move north or up in elevation if they are able; however, the pika already lives high in the mountains and lacks options to move higher up. Longer growing seasons might sound good for your garden, but they also benefit species like mosquitos and ragweed. Moose now suffer from parasites and ticks that winter temperatures are no longer cold enough to kill. Various species of bark beetles (which have killed millions of acres of conifers across the western US and Canada) now complete their entire life cycle in just one year instead of two (essentially doubling their population), aided by longer summers and winters no longer cold enough to kill them. Allergy-sufferers are due for increasing length and intensity of their misery seasons. Anglers in mountainous western Wyoming recently faced rivers closed to fishing because the waters were too warm and trout were stressed. Similarly, the sockeye salmon season was canceled in Bristol Bay, Alaska, in 2019 when river temperatures reached seventy-six degrees, too warm for the fish to swim up-river to spawn, and tens of thousands died.[27]

Human health is suffering as well because climate change alters the environmental determinants of health: water, food, and disease. Extreme heat is deadly; *Scientific American* noted that nine out of the ten deadliest heat waves ever have occurred since 2000, together killing about 129,000 people.[28] In addition, bacteria grow faster at warm temperatures, smog worsens (which affects asthma and heart disease), and heat can reduce crop yields and nutrition. Warmer temps also mean longer seasons for biting insects who also may transport diseases to new regions, such as ticks and mosquitos.

These and other examples of physical and ecological changes clearly show that climate change is happening *now*; it is not distant and remote, and it has cascading effects for human and nonhuman lives. Miami city streets flood with every high tide. In just three years, Louisiana had nine extreme flooding events—eight of the magnitude that are predicted to occur every 500 years, and one so big that it occurs every 1,000 years. For indigenous people in the far north, warming has upended their lives. The ice doesn't freeze thick enough or for long enough, making their travel and subsistence hunting more dangerous and less successful. Without ice, winter storms pound and erode their Arctic coastlines, forcing villages to relocate inland. Mosquitoes and biting flies emerge earlier and stay longer.

We've seen on the news that cities as well as forests are now vulnerable to fire (just in California: Los Angeles, Malibu, Napa, Santa Rosa, and Paradise) and that fire season is longer and more severe. The 2018 National Climate Assessment concluded that climate change played a greater role in increased wildfires than land management practices or past fire suppression. By 2017, the area of burned forest in the Pacific Northwest had increased by nearly 5,000 percent since the early 1970s.[29] The 2019–2020 fires in Australia, the continent's hottest and driest year on record, dwarf these numbers. Over forty-five million acres (an area roughly the size of North Dakota) burned in thousands of fires that killed well over a billion animals (and thirty-four human animals) and burned about 3,000 homes and several thousand buildings. Economists estimated the impact of the fires was about $2.4 billion.[30] Also in 2020, more than 3 million acres burned in the Arctic (exceeding all modern records), and almost 10 million acres burned across the United States (6 million of that on the West Coast). Wildfire scientists say that wildfires are behaving in unprecedented ways and that rising global temperatures and worsening droughts mean that the world has entered a new era of megafires.[31]

A difficult aspect of communicating climate change is that we cannot point to any one event and say climate change was its sole "cause." After all, we've always had fires, hurricanes, and droughts. Now, a scientific method called "extreme event attribution" can conclude (by historically examining probability and magnitude) whether an extreme weather event (like a heat wave, drought, or heavy rainfall) was made stronger and more frequent by climate change and the chances it would have occurred without it.[32] For example, two studies found that climate change greatly amplified Hurricane Harvey's record-breaking sixty-inch rainfall in Houston; researchers determined that climate change made the extreme rainfall three times more likely.[33] Researchers now conclude that climate change increased the risk of fire weather fivefold in the western United States and more than doubled the acres burned. Climate change made the bushfires in Australia 30 percent more likely. Increasingly, scientists are finding this "fingerprint" of climate change on extreme weather events worldwide, building a clear and solid case for our living in a climate-disrupted world. And it's happening within the 6 miles above your head.

WEATHER IS THE NEW POLAR BEAR

When researchers first asked people in 2003 what they most associated with "climate change" or "global warming," people said polar bears and melting ice. But in the past decade, the new symbol that people increasingly associate with climate change is the real-life, bizarre weather.[34] In surveys about climate change, people's mentions of weather events—killer heat waves, wild hurricanes and floods, huge wildfires—have quadrupled, and extreme weather is now one of the first associations people make with the climate crisis.

That makes sense. We hear about fires and floods across the country, and we witness "climate weirdness" in our backyards: a parade of record temperatures, winters that just aren't cold like they used to be, birds arriving earlier, and more intense yet sporadic rainfall. There is evidence that extreme weather may be helping break some climate silence. In North Carolina, all mention of climate change had been removed from state websites. But after hurricanes Michael and Florence blasted the state within twenty-three months, the governor created a Climate Change Interagency Council to strategize how to lower greenhouse gases emissions statewide.[35]

Some weathercasters are helping to change public opinion. The public has a high level of trust in local meteorologists, and TV weathercasters are good science communicators.[36] Surveys found that when people see weathercasters talk about climate change, their opinions change more rapidly and they perceive climate change as more personally relevant.[37] The survey also found that 95 percent of weathercasters now think climate change is happening, compared to just 54 percent who did in 2010. Yet over the past year, only one-third of weathercasters surveyed had informed viewers on-air about the local impacts of climate change.

For the last several years, the globe has repeatedly experienced too much water in some places (intense flooding from the US, India, and Japan) and the utter lack of it in others (fires racing through the Arctic Circle, Athens, and Australia). Climate change is a "threat multiplier," which has boosted the odds of record-breaking heat more than 80 percent across the globe.[38] These extreme events were precisely the kind of events scientists predicted would be more prevalent in a world impacted by climate change, but the scientists have been surprised by the speed, severity, and numbers.[39]

The events and images witnessed in the last few years are dramatic evidence that the climate crisis is here, right now. But that doesn't mean that it's easy to accept this evidence or know what to do with it. Even clear-cut scientific facts have communal lives beyond the science;[40] once released in the world, they obtain meaning through social interaction. We each, individually and collectively, must learn how to face, negotiate, and communicate the pragmatic and moral implications these facts have for our lives and for the larger Eairth. How do we face and acknowledge how such information makes us feel? How can we see ourselves as powerful social actors instead of paralyzed individuals? How can we bring this difficult topic into our social circles and talk through the contradictions and fears in a way that helps us engage more fully? How can we positively envision our lives (and our children's lives) going forward?

These are communal questions that require collective conversations. We often hear in the aftermath of disaster how people "came together." The accelerating changes and extreme events delivered by the climate crisis remind us that we are a community, and that we are connected to other humans and the entire living world—including the six miles above our heads in our precious, shared atmosphere.

COMMUNICATING A SUPER-WICKED PROBLEM

If you haven't read much about climate change or climate science before, that was a lot to take in. And now you might be wondering: Why is the most serious and comprehensive environmental problem in modern times not forefront and center in our conversations and daily activities? It's a good question with many complex answers that deserves to be examined.

Technically, the "solution" to climate change is straightforward and well-known: drastically reduce humans' combustion of fossil fuels. Although the solution is straightforward, getting there is anything but. The world is now highly dependent on fossil fuels, and our behaviors and ways of life developed around their use. And, the world's population is not united around the extreme urgency of reducing carbon emissions. Thus, the crisis of climate change isn't about the veracity of the science; the crisis is how we humans face and accept the problem and rapidly chart new ways forward to meet this significant challenge to all of our lives and the living world.

In addition, climate change is not like any other physical problem, environmental or otherwise. If there is a polluted river nearby, you can focus your attention and communication right there. You can see the problem, and perhaps smell the problem. Some government entity (perhaps several) are charged with protecting this river, of finding the source of the problem, and fixing it (though it still might take citizen pressure and/or lawsuits to prompt parties to act).

When scientists discovered in the 1980s a hole in the ozone layer high above the earth (in the stratosphere, the layer above the troposphere), it was straightforward, both to talk about and address. The metaphor of a "hole" (where dangerous solar radiation came through and caused cancer) resonated with people. The cause was known, and the solution straightforward: ban the use of CFCs (chlorofluorocarbons) in products (like hairspray) and appliances (like refrigerators). Alternatives to CFCs were available and fairly easy to switch out (although some use continues). (CFCs are also a potent greenhouse gas that remains in the atmosphere for decades, though the gas is far less abundant than carbon dioxide.)

Compared to the ozone hole or a polluted river, climate change is an abstract "wicked problem." That means it lacks simple or straightforward responses and has many interdependencies; it is socially complex and culturally embedded. And attempts to address it can lead to other unforeseen consequences.[41] Recently, a group of scholars labeled climate change a "super-wicked problem" because it possessed four features.[42] One, time is running out. (We need to take drastic steps in the next decade to avoid a several-degree global temperature rise.) Two, those who caused the problem also seek a solution. (That's us humans.) Three, central authority is weak or nonexistent. (No single entity is in charge of Eairth, although countries craft international climate treaties and apply pressure to carry them out.) Four, irrational "discounting" pushes responses into the future. This has both a social side and an economic side. We tend to think of climate change as distant and future, not something we need to act on now. While some people contend that addressing climate change would cost too much, the solutions are *exponentially less expensive* than the future impacts.[43]

Still other characteristics make climate change a bugger to talk about or act upon. We identify less with Eairth than Earth. Greenhouse gases are invisible, and you cannot look around and see climate change itself; only by comparison to past decades and centuries is change apparent.[44] If you do not pay close attention to when crocuses bloom or when the rainy season begins, changes over time are less noticeable. That makes it easy to assume that climate change is happening far away and to others, but it doesn't really affect me (even though if the Eairth isn't healthy, I'm not healthy). For this reason, the time-lapse images of glaciers melting and coral reefs dying by photographer James Balog are compelling visual records of change.[45]

Another difficulty is that public opinion is divided (far greater than two decades ago), not by knowledge about climate change but by *political affiliation*, which has confused policy and action with scientific findings. Even so, an analysis of public opinion by George Mason University and Yale

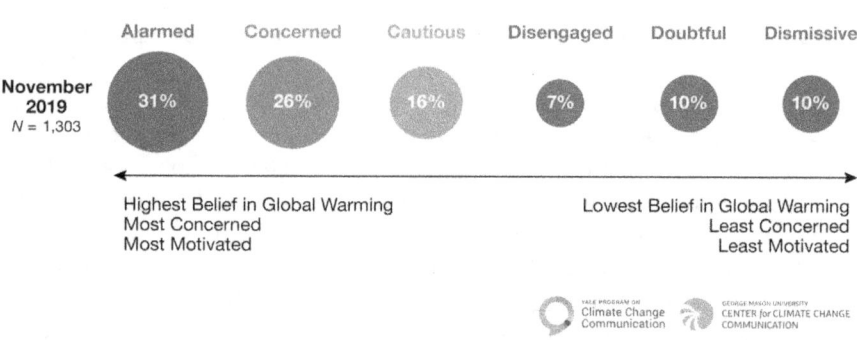

Figure 1.2 Global Warming's Six Americas.

University[46] has a bit of good news for communicating climate change. Researchers categorized survey respondents according to their level of belief, concern, and motivation about global warming into six distinct segments, which they call Global Warming's *Six Americas* (figure 1.2). In a recent survey,[47] the "dismissive" segment (who doesn't believe global warming is happening nor is human-caused) is just 10 percent. Although "dismissive" voices can be loud (even dominant) in news and political circles, they are a smaller percentage than many realize. Even the "doubtful" (10 percent) remain open to learning more, and some of the "disengaged" (7 percent) can be engaged. To me, this is an opening full of possibility for communication.

Additional good news is that public opinion is shifting in some countries. Shortly before the 6-million-strong worldwide "school climate strikes" in September 2019, residents surveyed in seven of eight countries called the climate breakdown the most important issue facing the world.[48] (Among US respondents, the climate crisis was third behind terrorism and affordable healthcare.) The majority of respondents said that they were already experiencing the climate crisis and didn't believe their respective governments were taking sufficient action. A survey of forty countries in June 2020 found that in the vast majority of countries, fewer than 3 percent said climate change was not serious at all.[49] Four of the five countries showing the highest levels of concern (85–90 percent) were from the global south.

NEW DIRECTIONS FOR CLIMATE COMMUNICATION

Communication is essential in our ability to address climate change, but it has struggled to break through the polarization and climate silence, which has created serious individual, social, political, and biospheric consequences. Even though climate change is present in our everyday lives, it too often remains (to borrow from *Harry Potter*) "that which cannot be named."

Literally. Florida banned the terms "climate change" and "global warming" in official state communication and reports.[50] Even after Hurricane Irma, Florida governor Rick Scott (later a US senator) said he was still unsure about climate change.[51] A regional National Park Service office told the New Bedford Whaling National Historic Park to remove references to climate change-related threats to the site such as flooding.[52] Avoidance is a not an effective way to address the most serious global problem of our time.[53]

Think for a moment of what you've read or seen about the climate crisis. Then consider whether more of that same type of information would move you to become engaged or active in the issue. Chances are, no. If you're like many people, you know climate change is serious and needs to be addressed. You might feel anxious or fearful about it, or conflicted because you rely on fossil fuels every day. When you read about worsening wildfires or collapsing ice sheets, you might emotionally tune out, thinking, "Those things are really horrible, but there's really nothing I can do about it."

A communicator knows that someone with a sentiment such as this is not likely to attend to messages, to engage, or to take action. More information or more science probably won't change that. An approach of "more" treats an audience's lack of information as the problem, when in fact, more information of the same kind can make a person's passivity or denial worse.[54] The problem is not insufficient information but more likely an inability to deal with the climate information we already have. This calls for new directions in communicating the climate crisis.

A traditional communication approach is to craft messages and frames (and cater to certain demographics) in ways that might influence individual attitudes and behaviors—a micro-level approach. Instead, this book views climate change communication as inherently about the interactions of individuals with all that surrounds them. Individuals are products of their interactions at the social level (friends and family, social media), the cultural level (pop culture and entertainment, cultural values and practices), and the macro level (the economic and political spheres with institutions like law and education, and the physical environment). In other words, an individual's behaviors are not chosen in a vacuum but are greatly influenced by how she interacts with and responds to communication at all of these levels in a holistic way.

Climate communication approaches would benefit from new strategies and new directions to better address this complex, multifaceted problem. Each chapter in this book presents a new direction or strategy to help climate change communication become more effective. The chapters bring in research and wisdom from a variety of disciplines (humanities, sociology, psychology, environmental studies, and others) to inform both individual understanding of the problem and broadened communication about it. Here is an overview of each chapter and its new direction for communicating the climate crisis.

As *chapter 1* has discussed, we live *within* Eairth's atmosphere and are greatly affected by what happens in it. Scientists agree that climate change is happening right now and is in no way a "normal" fluctuation. There is abundant evidence that it's human-caused and we have already witnessed many serious consequences in our climate-changed world.

Chapter 2 describes how the climate crisis evolved out of our burgeoning use of fossil fuels. All of us are embedded in fossil fuel culture whose fingers poke into every sphere, every product, and every arena of daily life. Consumerism is fossil fuel culture's defining value, shaping cultural identities and practices. Yet, climate change itself exerts enormous pressure for profound changes to our consumer identities, values, and daily actions. Addressing the climate crisis will take not just personal change in consumer behaviors and identities but also significant social change—a cultural transformation as well as an energy transformation. Social protest movements, reformist and radical, are explored for their social change potential. Communication plays a key role in social change: it can reinforce the fossil fuel status quo, or, communication can take the form of alternative narratives that disrupt existing practices and beliefs and challenge the current social order. Understanding how fossil fuel culture must change is a vital precursor to understanding how we as individuals can best contribute.

In light of fossil fuel culture, *chapter 3* explores the most effective role for individuals in fighting the climate crisis. The individualistic consumer identity fostered by fossil fuel culture tells us we can "fix" climate change with LED bulbs and reusable bags (or we feel shame for not taking enough consumer actions). Of course, minimizing one's impact on the Eairth is important and admirable! But these actions function best as a bridge to larger involvement because by themselves, voluntary consumer actions cannot address the systemic issues that created climate change. Instead, a shift from a role of individual consumer to a role of "social actor" is a powerful one. As social actors, individuals talk to and interact with social others (with whom they have strong ties and shared values), from a hiking club to a neighborhood association. When an individual acts as part of a collective, he moves from isolation to participation and creates vital social support for the social change needed to combat the climate crisis. A powerful social group could be a citizen group lobbying their representatives, a family marching in support of climate science, or a congregation that spreads the word about "creation care."

Even when you embrace your role as a social actor, climate change evokes powerful emotions. *Chapter 4* examines how to recognize and work through the emotions raised by the climate crisis. It's a simple fact that this huge issue can make all of us—climate acceptors, climate deniers, and everyone in-between—feel stressed and even paralyzed. At times, it might seem easier to ignore and "stuff" the worries and defenses you hold (unconsciously and consciously), which only heightens climate silence. People are unlikely to fully attend to communication about the climate crisis until they acknowledge and begin to process the difficult emotions—loss, fear, disappointment, grief, ambivalence—that they bring to this issue. The American Psychological Association recognizes that the mental health tolls of climate change are far-reaching and that "loss of place" is not a trivial thing. Building emotional resilience helps you adapt to stressful situations, self-regulate your emotions, and accept the face of the changed "normal."

Once you understand your emotional patterns and defenses and come to a place of emotional resilience, you're a social actor who is ready to talk about climate change. *Chapter 5* reveals that a key way to break the cultural silence about the climate crisis is simply talking about it with others. This elevates its importance throughout culture and helps us examine our feelings and experiences right here, right now. Conversation is a powerful tool to recognize and discuss emotions, find common ground, reduce polarization, and help change social and cultural norms. The chapter provides strategies and suggestions for entering conversations, including mock conversations with "friendly skeptics," which prepare you for climate conversations with family, friends, coworkers, and even strangers.

In light of the severe impacts of climate change, *chapter 6* brings the moral imperative of climate action to the forefront of the discussion, by asking, what moral responsibilities do we have for other humans and other species, now and for generations to come? One moral consideration is climate justice, which recognizes that those who contribute least to climate change are disproportionately affected and harmed: people of color, women, and the poor. The chapter also examines moral communication through the lens of religious worldviews. Even though all faith traditions have components in their teachings about "concern for the common Earth" and being stewards of creation, some religions (particularly those that merge religion and politics) actively deny the existence of climate change. Faith communities reach 85 percent of the world's population and are an important arena for morally based communication. So, what responsibility do privileged people and people of faith bear to talk about and act on climate change?

Negotiating a climate-disrupted world requires far more than switching to "cleaner" energy. *Chapter 7* discusses climate change as a symptom of a far larger crisis: an ecological crisis resulting from a human-centered valuation of—and separation from—the living world. To fully address the climate crisis, human belief systems must recognize our connections with Eairth systems. No matter our differing worldviews, we are embedded in a biosphere where all flourishing is mutual. Communicating a shifted worldview of humans' interdependency in the larger world—Eairth Citizenship, if you will—is a crucial accompaniment to climate action. The chapter also discusses a burgeoning "rights for nature movement" that codifies the intrinsic value and legal rights of the nonhuman world, from Lake Erie to the Te Awa Tupua in New Zealand.

Finally, *chapter 8* explores the growing realization that stories and language can effectively convey beliefs and spur engagement and action to address the climate crisis. Fully engaging hearts, hands, and minds spawns the creativity and imagination needed to fully separate from the "old story" of fossil fuel culture. This grandold story is a carrier of anthropocentric beliefs that sees humans as separate from and able to "control" nature. With climate change, we find ourselves in a blank chapter of transition from the old to "new stories" that will help direct us toward new ecologies and economies and new cultural narratives. Stories—told around dinner tables, conference rooms, and also in literature, film, art, music, performance, and comedy—are powerful. Storytellers can create for their listeners empathetic connections, experiential learning, story-consistent beliefs and behaviors, and less counter-arguing and resistance. New stories of interbeing with the living world can help us reckon with the past and envision the future, finding meaning and making sense of our lives amidst the larger, climate-changed world.

In case you've already thrown up your hands about climate change and believe it's too late to act, consider this. I possess what climate scientist and communicator Katharine Hayhoe calls "rational hope."[55] I accept that we are where we are, that it's bad, and that we must contend with the climate crisis for the rest of our lives. At the same time, I love life on this good Eairth—both on the crust and within the atmosphere—and I can truly envision positive, changed, and healthier ways of living here. Our choices *right now* matter a great deal in terms of how our future world looks and whether humans and the living world merely survive or thrive. And, I know that cultures are capable of profound, rapid change in times of crisis – and at the core of such change is communication.

NOTES

1. David Abram, *Becoming Animal: An Earthly Cosmology* (New York, NY: Pantheon Books, 2010).

2. This video describes what the sight meant to the Apollo crew: https://www.nationalgeographic.com/science/2018/12/earthrise-apollo-8-photo-at-50-how-it-changed-the-world/.

3. Contrary to all the songs about sunrise and sunset, the sun does not "rise" above a static Earth. Instead, the Earth revolves to face the sun once again. What looks like a sun rising over the horizon is actually the Earth rotating *down* and making the sun seem like it's going up.

4. "Special Report on the Ocean and Cryosphere in a Changing Climate," *IPCC*, September 2019, https://www.ipcc.ch/srocc/home/; Justin Gillis, "Should You Trust Climate Science? Maybe Eclipse is a Clue," *New York Times*, August 18, 2017, https://www.nytimes.com/2017/08/18/climate/should-you-trust-climate-science-maybe-the-eclipse-is-a-clue.html?mtrref=t.co.

5. Naomi Oreskes, "The Scientific Consensus on Climate Change: How Do We Know We're Not Wrong?," in *Climate Change: What It Means for Us, Our Children, and Our Grandchildren*, ed. J. F. C. Dimento and P. Doughman (Cambridge, MA: MIT Press, 2007).

6. *Mauna Loa Observatory in Hawaii*, https://www.co2.earth/daily-co2. See also: https:\climate.nasa.gov\vital-signs\carbon-dioxide\.

7. *Climate Science Special Report, Fourth National Climate Assessment (NCA4)*, November 3, 2017, https://science2017.globalchange.gov/.

8. https://www.cbsnews.com/news/arctic-hottest-temperature-ever/

9. Hannah Hoag and Jack Marley, "Climate Crisis—Here's What the Experts Recommend We Do," *The Conversation*, October 29, 2019, http://theconversation.com/climate-crisis-heres-what-the-experts-recommend-we-do-123238.

10. *Fourth National Climate Assessment (NCA4)*.

11. https://www.fueleconomy.gov/feg/contentIncludes/co2_inc.htm/.

12. *Earth on the Edge*, http://www.earthontheedge.com/how-much-co2-are-you-emitting/.

13. Oreskes, "The Scientific Consensus."

14. John Cook, Naomi Oreskes, Peter T. Doran, et al., "Consensus on Consensus: A Synthesis of Consensus Estimates on Human-Caused Global Warming," *Environmental Research Letters* 11, no. 4 (2016).

15. Kris M. Wilson, "Drought, Debate, and Uncertainty: Measuring Reporters' Knowledge and Ignorance about Climate Change," *Public Understanding of Science* 9, no. 1 (2000): 1–13.

16. Robert J. Brulle, "Institutionalizing Delay: Foundation Funding and the Creation of U.S. Climate Change Counter-Movement Organizations," *Climatic Change* 122, no. 4 (2014): 681–94; John Cook, Geoffrey Supran, Stephan Lewandowsky, Naomi Oreskes, and Edward Maibach, *America Misled: How the Fossil Fuel Industry Deliberately Misled Americans about Climate Change* (Fairfax, VA: George Mason University Center for Climate Change Communication, October 14, 2019), https://www.climatechangecommunication.org/america-misled/; Riley E. Dunlap, "Climate Change Skepticism and Denial: An Introduction," *American Behavioral Scientist* 57, no. 6 (2013): 691–98; *PBS Frontline*, "Climate of Doubt," http://video.pbs.org/video/2295533310/; *NASA*, https://climate.nasa.gov; *NOAA*, http://www.noaa.gov/climate; *UCAR*, https://scied.ucar.edu/climate; *Union Concerned Scientists*, https://www.ucsusa.org/.

17. Oreskes, "The Scientific Consensus," 76.

18. Naomi Klein, "Season of Smoke," *The Intercept*, September 9, 2017, https://theintercept.com/2017/09/09/in-a-summer-of-wildfires-and-hurricanes-my-son-asks-why-is-everything-going-wrong/.

19. Joe Romm, "The Real Fire and Fury Is in Greenland Right Now, and It's Super Scary. Scientists are Freaking Out," *Think Progress*, August 9, 2017, https://thinkprogress.org/a-rare-and-unusual-wildfire-has-hit-greenland-heres-why-thats-terrifying-2de794570798/.

20. Jim Steenburgh, *Secrets of the Greatest Snow on Earth: Weather, Avalanches, and Finding Good Powder in Utah's Wasatch Mountains and around the World* (Logan, UT: Utah State University Press, 2014).

21. LuAnn Dahlman, "Climate Change: Global Temperatures," *NOAA Climate.gov*, https://www.climate.gov/news-features/understanding-climate/climate-change-global-temperature.

22. Peter H. Kahn, Rachel L. Severson, and Jolina H. Ruckert, "The Human Relation with Nature and Technological Nature," *Current Directions in Psychological Science* 18, no. 1 (2009): 37–42.

23. *US National Phenology Network*, https://www.usanpn.org/.

24. Maya Kapoor, "Climate Change Is Unraveling Natural Cycles in the West," *High Country News*, May 11, 2017, http://www.hcn.org/articles/climate-change-is-disrupting-the-wests-spring-phenology.

25. S. J. Mayor, D. C. Schneider, J. Otegui, et al., "Increasing Phenological Asynchrony between Spring Green-up and Arrival of Migratory Birds," *Scientific Reports – Nature* 7, no. 1 (2017), doi:10.1038/s41598-017-02045-z.

26. Brian L. Sullivan, Jocelyn L. Aycrigg, Jessie H. Barry, et al., "The EBird Enterprise: An Integrated Approach to Development and Application of Citizen Science," *Biological Conservation* 169 (2014): 31–40.

27. Miranda Weiss, "Climate Change Has Setnetters Worried about Alaska's Sockeye," *High Country News*, June 1, 2020, https://www.hcn.org/issues/52.6/north-climate-change-has-setnetters-worried-about-alaskas-sockeye.

28. Gayathri Vaidyanathan, "Killer Heat Grows Hotter Around the World," *Scientific American*, August 6, 2015, https://www.scientificamerican.com/article/killer-heat-grows-hotter-around-the-world/.

29. Andrea Thompson, "Wildfire Season Is Scorching the West," *Climate Central*, July 28, 2017, http://www.climatecentral.org/news/western-wildfire-season-off-to-blazing-start-21661.

30. Center for Disaster Philanthropy, "2019–2020 Australian Bushfires," https://disasterphilanthropy.org/disaster/2019-australian-wildfires/.

31. Ed Struzik, "The Age of Megafires: The World Hits a Climate Tipping Point," *Yale E360*, September 17, 2020, https://e360.yale.edu/features/the-age-of-megafires-the-world-hits-a-climate-tipping-point.

32. Chelsea Harvey, "Scientists Can Now Blame Individual Natural Disasters on Climate Change," *Scientific American*, January 2, 2018, https://www.scientificamerican.com/article/scientists-can-now-blame-individual-natural-disasters-on-climate-change/.

33. Eve Andrews, "Climate Change Made Hurricane Harvey Wetter. Here's How We Know," *Grist*, March 2, 2018, https://www.youtube.com/watch?time_continue=7&v=bwtjqvHnD9s.

34. Shannon Osaka, "Move Over, Polar Bears: Climate Change has a New Symbol," *Salon*, July 30, 2018, https://www.salon.com/2018/07/30/move-over-polar-bears-climate-change-has-a-new-symbol_partner/.

35. John Murawski and Will Doran, "Cooper Sets Global Warming Goal to Cut NC Greenhouse Gas Emissions by 40 Percent," *Charlotte News Observer*, October 29, 2018, https://www.newsobserver.com/news/politics-government/state-politics/article220789175.html.

36. Pam Radtke Russell, "How TV Weathercasters Became the Unsung Heroes of the Climate Crisis," *The Guardian*, September 18, 2019, https://www.theguardian.com/environment/2019/sep/18/tv-weathercasters-shift-public-opinion-climate-crisis.

37. E. Maibach, D. Perkins, K. Timms, T. Meyers, and B. Woods Placky, "A 2017 National Survey of Broadcast Meteorologists," *Survey. George Mason University Center for Climate Change Communication*, March 1, 2017, https://www.climatechangecommunication.org/all/a-2017-national-survey-of-broadcast-meteorologists/.

38. Bala Rajaratnam, "Quantifying the Influence of Global Warming on Unprecedented Extreme Climate Events," *Proceedings of the National Academy of Sciences* 114, no. 19 (2017): 4881–86.

39. Andrew Freedman, "2018's Global Heat Wave Is So Pervasive, It's Surprising Scientists," *Axios*, July 27, 2018, https://www.axios.com/global-heat-wave-stuns-scientists-as-records-fall-4cad71d2-8567-411e-a3f6-0febaa19a847.html.

40. Candis Callison, *How Climate Change Comes to Matter: The Communal Life of Facts* (Durham, NC: Duke University Press, 2015).

41. Mike Hulme, *Why We Disagree About Climate Change* (Cambridge, UK: Cambridge University Press, 2009).

42. K. Levin, B. Cashore, S. Bernstein, and G. Auld, "Overcoming the Tragedy of Super Wicked Problems: Constraining Our Future Selves to Ameliorate Global Climate Change," *Policy Sciences* 45, no. 2 (2012): 123–52.

43. Climate- and weather-related events over the last five years directly cost the US more than $500 billion, according to the Federal Reserve. In addition to damage to natural resources and infrastructure, such events disrupt business and economic activity, affect the healthcare system, supply chains, agriculture, tourism, power generation, and labor productivity. If emissions continue to grow at historic rates, the losses will exceed the current gross domestic product of many states. Gina Heeb, "Climate Events Have Cost the US Economy More than $500 Billion over the Last 5 Years, Fed Official Says," *Markets Insider*, November 10, 2019, https://markets.businessinsider.com/news/stocks/climate-change-impact-on-economy-has-cost-500-billion-fed-2019-11-1028675379.

44. Adam Corner and Jamie Clarke, *Talking Climate: From Research to Practice in Public Engagement* (London: Palgrave, 2017); Hulme, *Why We Disagree*.

45. Photographer James Balog produced a documentary *Chasing Ice* about the retreat of glaciers. See this summary: http://www.ted.com/talks/james_balog_time_lapse_proof_of_extreme_ice_loss#.

Balog also documented coral death in "Chasing Coral," http://www.imdb.com/title/tt6333054/.

46. E. W. Maibach, A. Leiserowitz, C. Roser-Renouf, and C. K. Mertz, "Identifying Like-Minded Audiences for Global Warming Public Engagement Campaigns: An Audience Segmentation Analysis and Tool Development," *PloS One* 6, no. 3 (2011), doi:10.1371/journal.pone.0017571.

47. A. Leiserowitz, E. Maibach, S. Rosenthal, J. Kotcher, P. Bergquist, M. Ballew, M. Goldberg, and A. Gustafson, *Climate Change in the American Mind: November 2019* (New Haven, CT: Yale Program on Climate Change Communication, 2019).

48. Matthew Taylor, "Climate Crisis Seen as 'Most Important Issue' by Public, Poll Shows," *The Guardian*, September 18, 2019, https://www.theguardian.com/environment/2019/sep/18/climate-crisis-seen-as-most-important-issue-by-public-poll-shows.

49. Simge Andi, "How People Access News about Climate Change," *Reuters Institute for the Study of Journalism and University of Oxford*, June 16, 2020, http://www.digitalnewsreport.org/survey/2020/how-people-access-news-about-climate-change/.

50. Tristam Korten, "In Florida, Officials Ban Term 'Climate Change'," *Miami Herald*, March 8, 2017, http://www.miamiherald.com/news/state/florida/article12983720.html#storylink=cpy.

51. Marc Caputo, "Florida Governor Remains Unsure About Climate Change After Hurricane Irma," *Politico*, September 14, 2017, http://www.politico.com/states/florida/story/2017/09/14/florida-governor-remains-unsure-about-climate-change-after-hurricane-irma-114498.

52. Elizabeth Shogren, "National Park Officials were Told Climate Change was 'Sensitive'. So They Removed It from a Key Planning Report," *The Reveal*, September 10, 2018.

53. This is not to dismiss COVID-19, the pandemic which spread worldwide in 2020. But as we'll discuss, it arose because of human destruction of the environment, which stresses wild animals who are brought into closer contact with humans. In this sense, addressing COVID-19 and the climate crisis involve the same solution: protecting the health of the living world.

54. Lorraine Whitmarsh, Saffron O'Neill, and Irene Lorenzoni, eds., *Engaging the Public with Climate Change: Behaviour Change and Communication* (London: Earthscan, 2010).

55. Megan Kimble, "How Katharine Hayhoe Stays Hopeful as the Planet Warms," *The Texas Observer*, December 9, 2019, https://www.texasobserver.org/how-katharine-hayhoe-stays-hopeful-as-the-planet-warms/.

Chapter 2

Fossil Fuel Culture

"Talking to your speaker is totally normal now," I heard a commentator on the radio say. Siri, Alexa, Google Assistant, and others are your electric slaves, captive on coffee tables and countertops, listening 24-7 for your next command. "Like an ideal servant in a Victorian manor, Alexa hovers in the background, ready to do her master's bidding swiftly yet meticulously."[1] Some kids sincerely believe the speakers to be real women. The "personality teams" that designed Alexa to be smart and humble wanted to make consumption *frictionless*. As one author concluded, the existential threat of smart speakers is that they are meant to eliminate thought from our consumption and short-circuit the process of reflection that stands between a desire and its fulfillment in the marketplace.[2] Tech companies want "smart" devices in your home, office, and car to wire and colonize your personal space in the name of consumption. Increasingly, their "human" voices gain power and space in our emotional lives as well.

"Smart assistants" demonstrate how the filaments of fossil fuel energy now thread through so many aspects of our lives. The devices exemplify how our culture embraces ever new ways to use ever more fossil fuel energy, even though climate change has been in our cultural consciousness for more than thirty years. New gadgets appear and we buy them. Robotic vacuums cruise the hallways. Amazon delivers stuff in two days (often by airplane), and freight is the fastest growing source of greenhouse gas emissions.[3] Holiday lawn decorations require electricity to stay inflated (including the ironic polar bear near my house). Public bathrooms are increasingly an entirely electric experience: lights, toilets, water, soap, towels. At night, homes and apartments have a green glow from beaming gadgets that do not sleep. Restaurants heat the great outdoors with patio heaters. Open shop doors in all seasons waft out warmed or cooled air. Ironically, we love energy and the super-sized presence it plays in our lives, yet ascribe little value to it when we use and waste ever larger amounts of it. Abundant energy use is a symbol of wealth and privilege.

For better or worse, you and I are members of this culture that fossil fuels helped create. We were born into it, socialized into it, and fully participate in it. It's frankly impossible (now) to step free and clear of it. Modern society's relationship with fossil fuels is ultra-deep and all encompassing. The objects derived from (and run by) petroleum products mediate our relationships – to other humans, to other life, and to things.[4] We experience ourselves and daily life surrounded by—and *in*—the products of petroleum: lip balm, clothing, gasoline, dental polymers, electronics, and ubiquitous plastic, from tampon applicators to yogurt cups.

However, now that we know fossil fuels are changing the climate upon which all life depends, it's clear we need to transition to new energy sources, which will perforce change culture. It's a daunting task. It's not a simple matter of becoming "net zero carbon"; it's a highly complex matter of transforming a culture that was built around the use of carbon.

In *The Great Derangement: Climate Change and the Unthinkable*, Amitav Ghosh concludes that climate change is a crisis not of science but of *culture*—a culture he says revolves to a great degree around our desires for things, yet often conceals the matrix that brought them into being.[5] Given the extraordinary depth and breadth of fossil fuel culture and our very real attachments (physical and emotional) to the objects of it, one scholar asks whether we are "hopelessly attached to all these unsustainable pieces of nearly everything."[6] Modern life is replete with incredible comforts, conveniences, and achievements, and I am thankful for them. Attached, yes. Hopeless, no. A "culture"—that amalgamation of customs, objects, social actions, institutions, and achievements like the arts and knowledge—is continually co-created by its members, meaning that cultures change, along with their values, everyday practices, and relationships. Cultures have changed throughout human history, driven by various pressures and disruptions, including new energy sources.

For climate change communication, it's important to understand how our energy history is much entwined with who we are as humans and how we conceive of and represent psyche, body, society, and environment.[7] The history of fossil fuels is interlaced with significant cultural developments, which changed our world, eased our lives, and in the process, changed our relationship with nature. In fossil fuel culture, a defining value is consumerism, which profoundly shapes our identities. Communication plays a central role in the push-pull of change: climate change pressures change yet an established culture works to *avoid* change. Although current discourse largely supports the continuing inertia of fossil fuel culture, that discourse is being challenged by new social movement groups that seek substantial social change.

FOSSIL FUELS AND CULTURAL HISTORY

As you read these words, your body is running on sun energy. Energy is the capacity to do work, and life is all about transforming one form of energy into another: plants transform solar energy through photosynthesis into growth, which is transformed into kinetic energy in our bodies when we eat them. The plants are fueling you—allowing blood and air to circulate, muscles to operate, and brain cells to comprehend this sentence. Depending on season and location, you are likely reading with electric lights and staying warm (or cool) with a furnace (or air conditioner), most often powered by fossil fuel energy such as coal or natural gas. These fuels are themselves fossilized sunshine, created when plant matter decomposed hundreds of millions of years ago and compressed. The sun is the source of *all* energy—your personal kinetic energy, the food energy in your refrigerator, the gas energy in your furnace.

The history of fossil fuels began over 300 million years ago when the swampy forests in the Carboniferous epoch covered the earth. As science journalist Charles Mann explains, fungi had not yet developed to decompose plants and release their solar energy, so entire forests collapsed into the muck, decaying only partially into peat. Over centuries, layers of plants were crushed and heated, and peat became coal. At the same time, dead marine plants and organisms formed sticky layers of oil, gas, and other petroleum compounds. He writes, "In these smashed jungles and seabeds, glossy and black, solar energy waited, frozen in time, ready to be tapped."[8]

And tap we did. As Ralph Waldo Emerson wrote in 1860, "Every basket [of coal] is power and civilization, for coal is a portable climate."[9] Until coal was widely dug and transported, even the wealthiest of households were cold when temperatures dropped. With coal for heat, people could warm their entire houses, which made indoor plumbing possible. As Mann explains, life was utterly transformed.

> The impact of fossil fuels exhausts hyperbole. Take any variable of human well-being—longevity, nutrition, income, infant mortality, overall population—and draw a graph of its value over time. In almost every case it skitters along at a low level for thousands of years, then rises abruptly in the 18th and 19th centuries, as humans learn to wield the trapped solar energy in coal, oil, and natural gas.[10]

Fossil fuels influenced the design of our cities and houses, both of which were soon built with automobiles in mind. Affordable cars removed constraints on physical and social mobility for even the working class. Because burning coal produced noxious amounts of air pollution, when technologies developed to use oil (and later natural gas) for electricity production, the shift seemed so much more clean, efficient, and utterly modern. Technology and culture flourished; the Industrial Revolution, fueled by vast amounts of coal, forever changed the structure and nature of global relations in our search for and use of fossil fuels.

Over the centuries, our primary sources of energy have shifted from wood, slave energy, animals (horsepower, oxen power), wind (windmills and sails), and now to fossil fuel energy: coal, oil, natural gas, and refined products like gasoline. What fossil fuel energy provided that the other sources did not was "surplus" energy beyond what was needed for mere survival and that could be easily stored. Once society had this surplus energy, fossil fuel extraction, delivery, and use became a "keystone industry" for it kept all other industries going. Virtually all of our current infrastructure and products were designed with and for fossil fuels.

An energy system is woven into the economy, demographics, psyche, and political aspects of human society. The dominant source of energy operates via a long chain—the energy source, its converters, and related systems—that is welded together in a largely invisible way.[11] Converters include a ship's sail, an atomic reactor, or a power plant that burns fossil fuel to create electricity. Converters don't exist singularly but emerge and develop as part of "converter chains." For example, think of the chains of energy within agriculture: large-scale plowing and harvesting, transporting, storage (such as in silos), food manufacturing, refrigerated trucks, grocery coolers, and home refrigerators. An energy system includes not just the trucks, factories, and material objects but the social system, workforce, and institutions that support it.

Another example of a converter chain was the petro-chemical industry developed during and after World War II, when oil became "an agent of chemical and social metamorphosis."[12] Oil brought us plastics, pharmaceuticals, pesticides, and inks. It changed what people wore, viewed, and ate, and through cosmetics, drugs, and clothes, our bodies literally became oily.[13]

Thus continued a never-ending search to find more fossil fuels and to deliver them more easily. And it was always a search for more; despite their lethal consequences (deadly fogs from coal emissions, deaths from mining, explosions from drilling, spills, and air pollution), absent was a belief that we should use less. Though humans have feared running out of fossil fuels for over a century, new technologies help us find more, even if it means drilling a mile deep into the earth. As Mann concluded, "Far too often, we have been told we are facing crises of scarcity, when the deepest of our problems are due to abundance."[14] Former cries about Peak Oil (fearing plummeting supplies) have shifted to cries of Keep It In the Ground (recognizing that technology like fracking has boosted supplies of fossil fuel).

Our culture is now a fossil fuel-dependent one. It is the fishbowl in which we swim, making it difficult to recognize its ubiquitous and material presence. When I flip a light switch, I'm not thinking "coal enabled this light." When I eat a pineapple, I'm not thinking of the truck that delivered the crates from the field to the docks or the freighter that ferried my fruit from Costa Rica or New Zealand, or the garbage truck that hauls away the rinds (sadly, organic waste is rarely composted). On goes everyday life, and I'm mostly unaware of the long chains that power it. Petro-products order elements of my life to other aspects of society—from the furnace inside to street lights outside, from the corner gas station to the oil pipeline, from the federal subsidy for fracking to the state gas tax, from the stock market to the book written about it.

The ubiquitous yet taken-for-granted presence of fossil fuels in our lives is part of why climate change is a "super wicked"[15] problem: modern society developed the way it did because of the inexpensive, stored, surplus energy available in fossil fuels. Our dependency on fossil fuels is so encompassing that none of us can simply step free of it—right now. It's easy to understand why people struggle with the reality of climate change and at times want to retreat from it. To address climate change and the health of the planet, enormous and complicated social change is required, including a change of our identities.

FOSSIL FUEL VALUES AND CULTURAL IDENTITY

Imagine the exuberance that accompanied the electrification of America, the joy of indoor plumbing, and the plush comfort of central heating. Today, much of the world has the ability to feel unimpeded by the length of daylight or the temperature outdoors, and to live as though unbound by geography. The ability to "control" indoor lighting and temperature contributes to humans feeling somewhat free of nature's conditions.

Fossil fuels seemed to liberate societies of restrictive relationships to the land, such as land-based agriculture. Mechanical power changed the discourse of nature and "machine"; it was tempting to believe that humans truly dominated nature and were free of its constraints. (The costs of believing this was a multitude of machine-made nightmares, like the oppression of factory workers and the poor, and great harms to the environment.[16]) Thinking we were independent from nature deepened the imaginary gulf between "nature" and "culture." Believing humans are separate from and in control of nature is as dangerous as the actions undertaken under the guise of those beliefs.

Several cultural thinkers make the case that each dominant energy source is accompanied by particular cultural values, and as an energy source changes, so do cultural values. In "A Short History of Oil Cultures: Or, the Marriage of Catastrophe and Exuberance," Frederick Buell traces how energy history is significantly entwined with culture. Fossil fuels brought an age of exuberance, yet one increasingly haunted by catastrophe. The exuberance is currently manifested in new industries that seem to celebrate risk: nanotechnology, genetics, surgery robots, and driverless cars. Buell says such exuberance celebrates human supersession of nature and evolution, and breaches the boundaries between the human and the technological.[17]

For Buell, energy is the essential *prop* underlying our material and symbolic cultures.[18] *Symbolic culture* (entertainment, art, literature, and social interactions that attach meanings to objects and experiences) communicates what's deemed important to us as humans and as part of our lived experience. Alongside *material culture* (the physical objects and elements that surround us), symbolic culture illustrates how we understand (and question) what we value, what success and happiness mean, what love looks like, and how we view the beyond-human world. For Buell, an energy source such as oil is the "obsessive point of reference and clear determinant" over daily lives: "Together, oil and electricity wrapped people within their many infrastructures—roads, pipelines, telephone lines, power cables—even as it began doing something else of great cultural importance: reaching into and restructuring peoples' private worlds, identities, bodies, thoughts, sense of geography, emotions."[19]

One of the values shaped by fossil fuels, Buell concludes, is the defining Western faith in progress, a faith so strong that it is unthinkable that progress will disappear due to hardships, constraints, or climate change. Progress is perceived as linear—like climbing an endless staircase—and science and technology will us keep climbing, despite obvious limitations of capitalism's growth model for a finite, overtapped planet.

Profound faith in progress is rooted deeply in cultural identity. Thus, climate change poses a powerful threat to what we see as our way of life for it challenges the idea that the free pursuit of individual interests always leads to a general cultural and societal good. Increasing threats from climate change will—*whether we act or not*—drastically shake up our way of life and reorder the global distribution of power and wealth. Much of that power is derived from—and is closely related to—carbon emissions.

Ghosh argues that popular concepts like "adaptation" and "resiliency" slip easily into the progress metaphor because they don't lead to a different way of living or even reduced fossil fuel use.[20] Depending on its context, "adapting" to climate change doesn't require vastly different ways of being and living, at least for privileged people and countries.

In *Art and Energy: How Culture Changes*, art historian and social thinker Barry Lord examines the defining role a primary energy source plays in how cultures develop and change.[21] Lord posits that acquiring sufficient energy requires adopting certain values and priorities, while abandoning or suppressing others. Dependency on an energy source conditions us; when I power up my laptop, I am willing to ignore (or not fully consider) the coal emissions required to generate the electricity. When slave-energy was a major energy source (in ancient societies as well as in recent times), people in those societies had to mostly accept the cultural value that some people were able to be slave owners, while slaves had no control over their own lives or their children's lives.[22] Even if you did not own slaves, you benefitted from slave energy through the cost of goods and services. If you depend on an energy source, even if you disapprove of certain features or values of that energy system, it's hard to effectively question the system itself.

Another premise of Lord's argument is that the values and meanings attached to an energy source become a basic component of the value system of that culture. When the primary energy source was coal, it produced a culture of production (think steel mills) where workers strongly identified with their work and what they produced. Coal extraction is capital-intensive and requires a large, disciplined, and skilled workforce willing to do dangerous work underground, which instilled a strong work ethic and class consciousness. By contrast, when oil (and later natural gas) began to challenge coal as the dominant energy source in the 1930s, it promised immense reward for little investment and less labor to tap the fuel. The focus shifted from the site of production to the marketing of the product. Oil and gas are largely knowledge industries focused on demand, supply, and price, which fostered a culture of consumption instead of production. Instead of a coal-era worker's identity of "I am what I make," the cultural identity shifted with oil and gas to "I am what I buy."

Lord's final premise is that energy transition is an engine of culture change; if the dominant energy source changes, then the values and meanings at the base of that culture will change. He believes we are beginning a transition phase and ventures that the next culture will be one of stewardship (driven by renewable energy) where individuals will identify foremost with "I am what I save and protect."

That is hopeful news. Although Lord admits that oil and natural gas are unlikely to disappear entirely as energy sources, as they become less dominant, cultural values and identity will shift. A new cultural identity would see fossil fuel energy as old-fashioned and conservative. Because no energy source is value-neutral, renewables would prioritize new cultural values and suppress others. It is possible—and not good for Eairth—that consumption would continue to be a strong cultural value, even if driven by renewable energy.

No matter the degree or timing of the transition, it's important to consider the kind of culture that is possible—and desired—beyond our current one. As Buell noted, "We need to ask what we start finding when we cease living in oil as if it were our oxygen."[23]

CONSUMPTION: THE DEFINING VALUE OF FOSSIL FUEL CULTURE

On my laptop's calendar, Black Friday is now a holiday. Curiously, we now celebrate and honor consumption as much as our country's birthday and religious holidays. (And I cannot remove Black Friday from the calendar unless I remove all holidays.)

Perhaps the most significant product of fossil fuel-driven capitalism was modern consumerism.[24] Consumption is not merely a symptom of climate change; it is a primary root cause of it (along with population growth).

Fossil fuels are used to extract and produce consumer goods and ship their components and products all around the globe. It's easy for me to get excited about a $10 sweater and $20 pair of shoes because the cheap fossil fuels and even cheaper wages factored into the price are invisible to me – I'm not aware of the real cost to the people who made my sweater, and the environmental costs to land, water, and atmosphere.

Some critics have called consumerism "the religion" of the century.[25] Consumerism extends far beyond buying necessary goods to the fulfillment of a pseudo-spiritual life; we look to the marketplace to provide answers to our problems and a sense of meaning and identity (things once provided by religion). In *Enchanting a Disenchanted World*, sociologist George Ritzer said that "cathedrals of consumption" like modern shopping malls have much in common with religious centers, fulfilling people's need to connect with each other and God's creation.[26] As highly efficient, standardized, and interchangeable places, malls (and to an extent, religions) work to maintain enchantment in the face of increasing rationalization and disenchantment.

In *I Want That!*, Thomas Hine concluded that we now live our lives in the *buyosphere*, an all-powerful, encompassing, ubiquitous cultural arena where we are consumers in a marketplace rather than citizens in a civic sphere: "The buyosphere is not a civic space, but it is our chief arena for expression, the place where we learn most about who we are, both as a people and as individuals."[27] Try and think of a place or time where you are able to leave the buyosphere entirely; with gear and clothing labels, I even carry the buyosphere into wild nature. Hine says the buyosphere increasingly serves as the master logic of social relations; every constituent of culture—love, nature, sex, music—can be transformed into a commodity and subordinated to market logic with a price tag. Consider how an important relationship in your life might be different *without* the presence of discretionary consumer goods.

The design and advertisement of consumer goods is finely tuned. *Ultraconvenience* is key; an astounding 80 percent of our trash consists of items used just once—take-out containers, paper towels, plastic bottles of "energy" drinks.[28] This is *premeditated waste,* as is selling products designed to fall apart quickly, and the packaging itself (which can constitute 10 percent of an item's cost). Marketing successfully reframes "wants" as "needs," such as needing X to be a good parent or needing Y to be loved, popular, and beautiful. Marketing nurtures consumer dissatisfaction more than it does satisfaction. No sooner than I buy a refrigerator or a camera, an updated version appears in the marketplace; *dynamic obsolescence* is a strategy to quickly outdate a product and create a "need" for the new and "improved."[29]

Each of us has been trained as a consumer since the age of three, which is when advertisements began to target you. Each day you receive thousands of ads persuading you to buy; each day you receive virtually no messages asking you not to buy. Essentially, "marketing forms a kind of wallpaper to our lives."[30]

When you encounter an ad, its primary appeal (no matter the product) is to the emotional portion of your brain where fear, pleasure, and desire work together to encourage consumption.[31] Advertisers seek to trigger your "wanting" something by tying the item to the pleasure it will produce, and also to trigger your fears that not buying it will make you less safe, less happy, and less in control. When fears or threats arise, you can become very attached to objects that seem to lessen those fears. Constant triggering of your attachment system by external threats can develop an anxious attachment style, which can lock you into habitual purchases or behaviors and tap into your deepest anxieties, particularly social status.

For example, think of the habitual behaviors and fears attached to your cell phone and the social media you access with it. By one estimate, phone-attached users touch or tap their phones 2,600 times a day.[32] Anxious attachment conditions you to seek proximity to and ownership of consumer goods as a refuge from fear, even though the fear is often quite unrelated to the product itself. In many ways, we operate within a general culture of fear and marketing heightens this fear, keeping our emotional brains in a constant state of vigilance and putting high value on anything that looks like a safe refuge. Fears of climate change can even trigger attachment and push consumers to greater levels of consumption as a way of soothing the uncomfortable feelings that arise.[33] Constant marketing messages can tip the balance toward an "endless cycle of consumption with no real pleasure beyond the momentary buzz of the acquisition."[34] It's hard to find satisfaction or satiation with our purchases, and often instead, there's a greatly diminished utility (or pleasure) with each additional purchase.

A man once told me he thought humans are born to be consumers and we always would be, as though consuming were an innate part of our DNA (not just to fulfill basic needs). That's the strength of consumer identity in fossil fuel culture: "I am what I buy." If a major part of how you perceive yourself as a person is as a consumer, you may view climate change (even subconsciously) as a great threat to that identity. How might you reframe this?

First, if you think that climate change means you have to give up consumer products, reframe that to recognize what you are *already* losing and sacrificing. Increasingly, we all know someone who was affected or displaced by an extreme climate-related event, someone who lost possessions, livelihood, security. The climate crisis threatens more human and animal life than *any war* we've ever faced. So perhaps it does warrant a bit of sacrifice; widespread rationing during World War II was difficult but widely accepted as being for the common good.[35] Every product—widescreen TV or new jeans—is embodied with "sacrifices" of earth elements, energy, and workers' lives, regardless of whether you are aware of them.

Another strategy is to be more accepting of change. The only constant in life is change, and climate change is accelerating change and reshaping human life on the planet, including the choices we make in the marketplace. If the price of goods and services reflected the true cost of carbon and earth elements, yes, many would be more expensive. Rosemary Randall, director of Cambridge Carbon Footprint, produced a tangible list of probable changes to our everyday lives when we address climate change.[36] First, we would spend more on essentials like housing and food, and less on nonessentials like clothing, holidays, and entertainment. Randall predicts that some familiar jobs would vanish, although new jobs would appear (which is already happening, such as solar installers). And, travel would change, whether that means flights cost more or most of us would travel less. Randall imagines that some current "freedoms" would be curtailed, perhaps the freedom to buy a gas-powered car or drive into the downtown core. If you're not being "solution averse," how might these changes bring about some very *positive* outcomes? Personally, I would look forward to cleaner air and fewer health problems from pollution.

Another reframing is to see changes as added positives and not just losses. Imagine more simplicity and space at home with fewer goods. For appliances and electronics, imagine them being built to last and easy to update and repair with pop-in components and solar charging.

Finally, shift your participation in the marketplace from frenetic and mindless to considered and mindful. A young woman admitted to me that she used the web browser "Dash" button to order Pop-tarts online, rather than walk to the corner convenience store, an example of how electronics can make consumption frictionless. *Mindfulness* can help break the hold that marketing exerts on your emotional brain. Contemplate rather than react; take a deep breath and reflect before you enter a store or press Place Order. Observe what triggers you—convenience, price, habit, or unconscious emotions like self-esteem or fear. Consider, of all your possessions, what do you truly value and have a *relationship* with?

To successfully transition beyond fossil fuel culture, the meanings we place on consumer goods and their role in our identities will evolve beyond "I am what I buy."

CLIMATE CHANGE, SOCIAL ORDER, AND CULTURAL TRAUMA

Some days, I feel hopeful that fossil fuel culture is changing: international treaties, protests against energy developments, more renewable energy, lawsuits, and divestment from fossil fuel stocks by a thousand institutions. Every few weeks, a large company announces its goals for zero-net-energy in twenty to thirty-five years or even net-negative emissions. Climate Action Network, an umbrella group for 1,300 NGOs (nongovernmental organizations) in more than 120 countries, noted the wide range of people taking action: youth groups, faith groups, indigenous peoples, health workers, farmers, trade unions, and some financial institutions.[37] The Pentagon recently called climate change a national security concern.[38] These are very positive things, of course, but the fact remains that greenhouse gas emissions have not decreased worldwide, and climate change has not become crucial for many people. Drastic action is urgent: a report by the UN Intergovernmental Panel on Climate Change said that by 2030 we need to be on a path for CO_2 emissions to fall by 45 percent to keep global temperatures from rising more than 2.7 degrees Fahrenheit. This is a huge reduction in a very short time frame; how can fossil fuel culture be transformed?

Throughout human history, cultures have changed, sometimes dramatically and sometimes prompted by a new source of energy and/or technology, and sometimes by ecosystem or social collapse.[39] But just as often, cultures work to *avoid* change and maintain the status quo, preserving existing power relationships and institutional structures.

Challengers of the status quo have always faced an uphill battle. In the 1500s, Niccolo Machiavelli wrote in his classic *The Prince* of the arduous task of social change:

> There is nothing more difficult to carry out, nor more doubtful of success, nor more dangerous to handle, than to initiate a new order of things. For the reformer

has enemies in all those who profit by the old order, and only lukewarm defenders in all those who would profit by the new order. This lukewarmness [arises] partly from . . . the incredulity of mankind, who do not truly believe in anything new until they have had actual experience of it.[40]

Seven hundred years later, Machiavelli's observations still have wisdom. Those who profit by the "old order" include the powerful fossil fuel energy industry (which donates millions to politicians), and those who benefit from inexpensive fossil fuels that power production and distribution, which includes every one of us. True cultural change is not the simple result of new laws or protests but complex interaction and transformation at all levels of society.

The climate crisis presents an extremely significant challenge to what sociologists call the *social order*, or the ways in which the components of society work together to maintain the status quo and remain stable. As an individual, you orient yourself to the narrative of the society into which you are socialized and undertake a fairly unconscious repertoire of behaviors and social interactions.[41] You drive to work. You pick up take-out for dinner. You turn on the TV. Your daily activities are embedded in the world around you – from technologies and the built environment to other humans and institutions. You internalize how fossil fuels structure your life and shape your identity: cars seem essential to participate in social events, the workplace, and marketplace. You accept and participate in the social order, and life seems predictable.

You learn from social settings and institutions what are normal and routine interactions: the power relationships in a workplace, and the appropriate behavior expected in a bank, a shop, a classroom. Institutions (including education, political representation and governance, law, economy, religion, and the marketplace) follow rules of interactions that continually reproduce the social order. Collectively, we accept the given social order as morally appropriate and normal, and we rely on it to see, respond, and attach meaning to the social processes unfolding around us. I see gas prices going down, I witness inaction by my leaders on climate change, and I observe the news media and police response to climate protests; I might conclude from these is that climate change is not a crisis. The dominant belief system of the social order is highly anthropocentric (human-centered and dominant over nature) and supports our energy-hungry lifestyles over the costs to ecosystems and other species.

Despite the change pressure exerted by the climate crisis (and a growing number of individuals and institutions), what exists right now is *social inertia*; humans are taking few actions, the status quo of fossil fuel culture continues, and the climate threat hyper-escalates. Sociologists Robert Brulle and Kari Norgaard say that social inertia occurs when powerful, interrelated cultural, institutional, and individual processes work collectively to inhibit action and social change.[42] They claim that what keeps us stuck in social inertia over climate change includes institutional conflicts over policies, attacks on climate science, development of denialist ideologies, and our own individual responses.

Large fossil fuel companies work hard to maintain social inertia. Even though Exxon Mobil's own scientists warned in the 1980s about impending "catastrophic" global warming,[43] the company plans to increase its production by 35 percent by 2030.[44] At least a dozen oil giants (including Shell, BP, and Chevron) are planning not an orderly decline in production but instead on ramping it up. Over the last thirty years, five major oil corporations spent $3.6 billion on ads that often *greenwash*—boast a "green" climate change image that doesn't match overall performance.[45] Greenwashing through advertising has been a primary strategy to manipulate environmental discourse and influence political outcomes around climate change, which greatly reinforces social inertia.

In your lifetime, you have witnessed shifts in the social order, such as the increasing presence of technology that shifted social interactions among individuals and institutions: think of the effect of internet search engines on the library reference desk, and Siri's effect on you and the marketplace. Yet these and other shifts have not fundamentally altered the moral appropriateness of fossil fuel culture nor the dominant human-centered worldview. If anything, increasing technology slides easily into fossil fuel culture, for it requires more energy to create and continually upgrade.

But the social order can be disrupted in such an enormous way that it profoundly interrupts or challenges routine ways of acting and thinking. Disruption began when climate scientists published (and the media publicized) more and more evidence that human actions were warming the earth and creating extreme physical changes with widespread and severe impacts. The narrative emerging from climate science challenged existing global economic structures and clashed with beliefs about humans' dominance over the living world.

Such a narrative can destabilize the social order. At all levels—individual, institutional, cultural—a new narrative questions ordinary ways of interacting. Climate change also challenges "ecological habitus," or each person's *ecologically relevant* yet taken-for-granted everyday practices[46] (eating strawberries in January, watering a lawn, buying wine from New Zealand, driving to the mall). When we become aware of climate change, it clashes with our everyday carbon-producing activities, which creates anxiety and threatens security.

When the social order is profoundly destabilized, conflicts and questioning can lead to *cultural trauma*, a systematic disruption to the way society accepts and sanctions the current social order.⁴⁷ Cultural trauma occurs when the existing social reality that anchors our personal identities and social interactions is significantly challenged or displaced. A cultural trauma shifts how we perceive our collective understanding (such as what climate change means for life moving forward). The dislocation may occur internally (in our collective consciousness) or from prolonged and cumulative external events like Hurricane Katrina in New Orleans and Hurricane Maria in Puerto Rico (which both greatly undermined citizens' expectations of government protection).⁴⁸ During both internal and external dislocations, new alternative narratives emerge that challenge the existing order and point to its inability to respond. Sociologists maintain that widespread and profound cultural trauma is needed to create spaces and reasons for significant change; mounting climate change evidence alone may be insufficient.

The COVID-19 pandemic was a prime example of a cultural trauma to the social order. It profoundly disrupted daily routines and social interactions, including to economic, institutional, and political spheres. Closing schools and workplaces created widespread ripples—for healthcare and first responders but also for food producers and distributors, restaurants, day care, transportation, retail, and entertainment. Shutdowns cleared skies, but crashed the stock market. (If people recognized the virus's common roots with climate change—environmental destruction that created health emergencies—the cultural trauma of the pandemic could serve as a lever to significant social change that would benefit both problems.)

At the personal level, internal and external disturbances are unsettling to our personalities. "Cultural trauma undermines an individual's sense of security and leads to a destabilization of the self, both of which provoke emotions including anxiety, distrust, pessimism, and insecurity."⁴⁹ What are possible responses in the face of cultural trauma?⁵⁰ One response to climate change is to retreat – to become passive and marginalize or discount it. A second common reaction is "ritualism" or continuing old behaviors and ignoring the now-known implications of them. A third response is rebellion, which attempts to alleviate trauma by attacking the new questioning narratives and all that they imply.⁵¹ To move beyond these typical responses requires that we individually and collectively reconstruct self-perceptions and actions.

At the institutional level, cultural trauma destabilizes the governing process. As trauma increases, environmental concerns may be seen as more relevant and interjected into institutional behavior norms ("the way we do business" or "the way we protect the public") but there is often conflicting pressure about how to respond. (For example, some cities and branches of the military readily acknowledge that climate change is altering their operations. But many other institutions are continuing "business as usual" and have not acknowledged the relevancy of climate change to the way they operate.)

This struggle creates confusion and disputes, but it can also lead to new and innovative institutional practices. Institutional conflicts have included permitting fossil fuel pipelines, tribal rights over energy production, and piecemeal efforts to curb emissions (such as California's emission controls that are stronger than federal standards).

At the cultural level, the climate crisis challenges beliefs about neoliberal capitalism and the role of government. Response to this challenge thus far has been to either maintain the status quo or attempt modest reforms (such as market incentives to price carbon emissions) and new technology. Such responses may be "socially acceptable narratives that maintain the existing social and institutional order" and gradually transform energy systems, but climate scientists note that the pace of this reform is inadequate to address climate change risks.[52]

If climate change threats were fully acknowledged by all levels, climate change would become a major, recognized cultural trauma[53] in clear need of resolution and change. But for now, too much social inertia prevails.

To move toward a cultural transformation that would successfully address the climate crisis, cultural trauma would be used as a *lever* that forced social change instead of social inertia. Normal practices must be questioned and disrupted in ways that affect and alter routines at all levels. This could involve clashes between those socialized in the old social order and those seeking a new social order (such as climate activists demonstrating worldwide).

Transformation could also include bold actions at the institutional level that disrupt routine practice and challenge collective identity.[54] The Green New Deal (first proposed in the UK and later introduced by US congressional Democrats) would boost "green" jobs and renewable energy; it was a bold action that faced stiff opposition from the social order. Transformation could involve new organizational practices and even new organizations that modify social interactions and technology in ways that filter down to individual practice (such as rural electric cooperatives that switch to solar energy). Various social imaginaries can help individuals envision self-reflexive transformations of their identities, worldviews, and practices.

An example of community level transformation is the Transition Town Movement, which sees itself as both a social movement and a grassroots innovation for sustainable, resilient living at a relocalized level.[55] When the initiative began in the UK, it was based on four assumptions: that it is better to plan ahead for life with dramatically lower energy consumption and climate change, that communities lack resilience in this regard, that people need to act collectively now, and that the collective genius of a connected community is better able to build new ways of living on the planet.[56] Over 1,200 such transition town initiatives exist around the globe.[57]

Although it sounds uncomfortable, cultural transformation requires a severe dislocation of existing social practice[58] across society. But as Brulle and Norgaard point out, climate change itself will create severe dislocation *whether or not* we take action. If climate change proceeds unabated on the path toward a projected global temperature increase of 4 degrees Celsius (7.2 degrees Fahrenheit) by 2100, this is "incompatible with any reasonable characterization of an organized, equitable, and civilized global community."[59] The other path—mitigating carbon emissions and limiting temperature rise—also requires fundamental, profound changes in economics and lifestyles. Either way, "Western societies will simply have to come to terms, collectively, with ways of living which differ radically from those that they have become accustomed to."[60]

COVID-19 gave us a taste of what radically different daily life might look and feel like. Widespread working from home (if you were lucky), massive unemployment, and simple elements of daily life altered or curtailed. People cooked, took walks, and connected with family and friends, virtually or otherwise. Amidst the widespread suffering and devastating death toll, many people managed to practice resiliency in the face of massive disruption of daily routines and to support disadvantaged others.

COMMUNICATION AND THE CLIMATE CHANGE MOVEMENT

Communication scholar James Carey called communication a symbolic process whereby reality is produced, maintained, repaired, and transformed.[61] Just as communication has helped produce and maintain the current social order that heightens the climate crisis, it likewise can help transform it into a new social order and culture. As noted above, sociologists say that what's needed to break social inertia are new narratives that greatly disrupt routine practice and challenge collective identity at all levels of society. The narratives must be destabilizing and widespread to lever cultural trauma into drastic social change.

For environmental change, many people look first to the environmental social movement, which arose in the 1960s and 1970s alongside antiwar, civil rights, and women's rights movements. It was a time of questioning and challenging authority, and demanding that governments legislate better treatment of air, water, land, and people. A social movement operates outside institutional and governing structures to bring about social change. Individual social movement groups (varying in organization, money, and formal structure) operate within a larger movement to achieve an overarching common goal: action to help the environment and/or climate change. Movement groups can mobilize citizens at the "grassroots" level (growing from the bottom up), although group tactics and targets vary greatly. Some social movement groups form around local issues such as environmental racism and toxics.[62]

For most large, long-established environmental movement groups like the Sierra Club, Greenpeace, Natural Resources Defense Council, and Environmental Defense Fund, climate change is just one of many issues they address. Now, hundreds of newer groups around the world focus exclusively on climate change (such as 350.org and Climate Action Network). A common focus of the environmental (and climate) movement—then and now—has been policy and regulation. Despite some successes (especially with grassroots support) environmental groups have not been able by themselves to break social inertia and produce significant climate action.

Climate change movement groups can be roughly categorized as *reformist* or *radical*, and the majority are reformist. Reformist groups seek changes from *within* existing market and governing structures (such as carbon taxes and emissions reductions), using *institutional actions* such as lawsuits, testifying, and lobbying. Reformist standpoints (which some sociologists categorize as ecological modernization or green governmentality[63]) maintain that the economy benefits from moves toward environmental quality; in this case, growth-oriented capitalism determines acceptable environmental regulations. As one critic said, "ecological modernization is essentially a discourse to ensure economic growth and to co-opt industrialism's environmental critics in the form of a managerial rhetoric."[64] Reformist groups don't typically seek to transform existing structures and institutions, which can contribute to social inertia.

Some movement groups have moved far from the "grassroots" of citizen involvement, operating primarily as independent NGOs where staff members direct tactics toward institutions (policy, technology, the market, international climate treaties).[65] The largest environmental and climate groups possess great resources, but their highly bureaucratic structures can make them less successful and more reformist in their actions.[66] Many environmental groups appeal to a limited demographic and struggle to move beyond an "environmentalist" stereotype.

A *radical* approach recognizes very real ecological limitations first and foremost—limitations that should transform how the market and governance operate, resulting in greater environmental protection. Unlike reformist actions, radical actions take place *outside* the systems and institutions they seek to change. The masses of people around the world taking to the streets over the climate crisis are engaged in nonviolent civil disobedience protests – a kind of *direct action* meant to question taken-for-granted practices; disrupt and alter daily routines; and draw the attention of the larger public, news media, and corridors of power. Direct action protests contribute appreciably to a *participation orientation* at the grassroots level with citizens demanding significant change from the status quo. In contrast, a *power orientation* views change as emanating from leaders and policy (which for climate change has not happened). For cultural transformation, both orientations are necessary to successfully respond to the climate crisis.

"Radical" comes from a word meaning "proceeding from the root," which is *not* the same as a specific place on a political spectrum. Climate change and climate justice groups are radical when they seek to change the root causes of climate change: dependence on fossil fuels and an anthropocentric worldview. In a sense, Buckminster Fuller was describing radical change when he said, "You can never change things by fighting the existing reality. To change something, build a new model that makes the existing model obsolete." Radical climate change groups (including Peaceful Uprising, Rising Tide, Climate Justice Alliance, and many others) use language such as "system change, not climate change" and "climate emergency" to mobilize broad participation and garner attention through fear for the future and the sentiment of crisis.

Radical groups maintain that conventional campaigns by reformist groups do not work: you can't overcome entrenched power with persuasion and information that seeks gradual reform and compromise.[67] Thirty years of inaction on the climate crisis and continually rising carbon emissions seem to support their conclusion. Historically, civil resistance models have succeeded. Lunch-counter and bus sit-ins during the civil rights movement were disruptive: protestors broke segregation laws, created economic costs (by police and businesses), attracted the attention of media and seats of power, and eventually obtained new laws (though the 2020 protests against systemic racism made clear that new laws alone did not create sufficient radical change).

The *Wall Street Journal* said that protests (about various issues) in 2019 rattled governments around the world, noting that the people who poured into the streets often shared tactics and slogans, expressed a similar mix of social and economic discord, and demanded sweeping changes to the existing political order.[68] The article referred broadly to protests (which were sometimes violent) in Hong Kong (about encroachment of China's central authority), Barcelona (about Catalan independence), Chile, Bolivia, and Lebanon. Researchers have also noted connections between changing climatic conditions and acts of antigovernment discord and violence, such as in Syria.[69] Almost all climate protests—which demand radical change in the social order—have been nonviolent.

Millions of protestors have already marched for climate change. A nationwide poll found that nearly four in ten Americans personally identify with climate activists, and about half support climate activists who urge elected officials to take actions to reduce global warming.[70]

This raises the question: how many activists and how many protests does it take to effect social change? Some interesting research shows that the answer is actually a fairly small critical mass. Erica Chenoweth, a political science professor at Harvard University, analyzed hundreds of nonviolent campaigns across the globe from 1900 to 2006.[71] She found that it takes roughly 3.5 percent of the country's population actively participating in civil protests to cause real political change; there weren't any campaigns that failed if they achieved this threshold of participation.[72] Strength in numbers causes severe disruption that paralyzes normal life and societal functioning. In the face of such disruption, it becomes harder to justify the status quo, which can lead to significant change.

The "*3.5 percent rule*" is good news. Rather than trying to influence climate deniers or doubters, a climate movement can seek participation from a broad cross-section of "alarmed" and "concerned" individuals. The power of 3.5 percent lies in the signal delivered by a critical mass of diverse people peacefully protesting to the power structure and fellow citizens. With sizable participation, the movement is more likely to win support from police (and/or military) and more likely to pose a threat to entrenched authorities or industries.[73] Massive protests mold opinion and shift loyalties. With nonviolent, civil disobedience protests, there are few barriers to participation—protesters represent all ages and all walks of life—and thus demonstrate broad commitment and a broad demographic (beyond traditional environmentalism).

Another way to examine the power and effectiveness of direct action protest is to contrast its success even when public opinion lags far behind. During the Civil Rights movement, a 1961 Gallup Poll found that only 28 percent of US respondents approved of the lunch counter sit-ins and freedom buses. And, when the US Supreme Court supported same-sex marriage in its 2015 ruling, only 57 percent of the public approved of it.

Let's take a look at two of the biggest direct action protest groups on climate change: Extinction Rebellion and the youth climate activists. Both belong to the digital age with a kind of "hyperspeed," taking advantage of technology and the internet to quickly share information and marshal citizens and/or members to gather and undertake specific actions.[74] Encrypted-messaging software such as WhatsApp and Telegram allow large numbers of protestors (who have never met each other) to communicate anonymously.

Extinction Rebellion

The civil resistance model of Extinction Rebellion (XR for short) claims that vested interests only change when forced to do so: "The rich and powerful are making too much money from our present suicidal course. You cannot overcome such entrenched power by persuasion and information. You can only do it by disruption."[75] This is an alternative narrative.

XR quickly made a name for itself for their peaceful, mass civil disobedience protests, first in the UK where they blocked a bridge over the Thames for a week, glued themselves to buildings, and chained themselves to a company headquarters.[76] Weeklong uprisings undertaken simultaneously in five cities were designed to annoy and disrupt. XR made their protests safe and welcoming for families and is committed to nonviolence.

XR's demands and language are clear and well-crafted (and now compiled in a handbook).[77] XR's first demand is "Tell the truth." (The handbook quotes George Orwell: "In times of universal deceit, telling the truth is a revolutionary act.") Second, the government must act now. And third, the government must be led by the decisions of a representative Citizens Assembly on climate change and ecological justice. Protest banners call for "Frugality, Humility, Empathy." XR also makes sure to openly acknowledge the often paralyzing emotions about climate change and the necessity to grieve for what is being lost.

XR's goal is to draw a critical mass of the British population (3.5 percent or 2.3 million) to actively participate in their protests. XR reports close to 500 affiliate groups in seventy-five countries, including in New York and Los Angeles (3.5 percent in the US equals 11 million).

In 2019, XR protestors super-glued themselves to each other and the passageway of the US House of Representatives to disrupt votes and demand passage of a resolution declaring a "climate emergency."[78] Using alarming images and messages, XR protestors urge action now, challenging government targets as being too slow and too low.[79] In a funeral march for mass extinction in Germany, campaigners dressed in black mourning clothes and carried a coffin. Inside the EU headquarters in Brussels, protestors staged a "die in" and covered the floor. In front of the *New York Times* building, protestors blocked traffic. Members spilled fake blood down the steps of the Trocadero Esplanade in Paris. Protesters blocked the doors of a shopping mall in Paris and buried their heads in the sand (literally) on a beach in Sydney. XR said it wants to create economic disruption to halt destruction of the planet and prevent catastrophic climate change, for which participants are willing to go to jail.

Do such tactics work? One investigative media site concluded that XR's language of urgency is shaping the political agenda.[80] During nine days in April 2019, more than 1,000 XR demonstrators in London were arrested.[81] Protestors came from all walks of life: doctor, poet, engineer, shop worker, student, security worker, manager. A journalist for *The New Yorker* reported that XR achieved two of their demands in the UK, passage of a climate emergency declaration by members of Parliament, and formation of a citizens' assembly.[82]

Youth Climate Movement

It began in August 2018 as one teenager, Greta Thunberg (pronounced TOON-berry with a roll of the R's), sat each Friday on the steps of the Swedish Parliament with a protest sign. By 2019, upwards of 10 million youth on the streets in 150 countries demanded action on climate change. This is truly a grassroots, direct action movement. Thunberg asked fellow teens to join School Strikes for Climate Action (also known as Fridays for Future); "I'm striking from school to protest inaction on climate change—you should too."[83] Many young people were spurred to action by photos and videos circulating on social media from their icon Thunberg. Strikes spread to every continent except Antarctica. Similar (and often affiliated) youth movement groups include the Sunrise Movement, Our Time, and Zero Hour.

The social change message of youth activists is compelling. Young people are extremely worried about their future and consider climate change a matter of intergenerational justice: adults are failing youth by failing to take action. The messengers at the youth protests have flipped the way they are seen; rather than adults teaching and supervising them, the youth are telling adults what they must do: "If you won't act like adults, we will."

Unlike mainstream environmentalism, participants in the youth protests better represent the globe's diverse population. As one sociologist noted, the tens of thousands participating at a Washington, DC march September 20, 2019 (a day of protests worldwide) were young, racially diverse, and overwhelmingly female.[84] Sociologist Dana Fisher surveyed 100 organizers and 200 participants of that Friday protest and found that more than a third were people of color and 58 percent of participants (and 68 percent of organizers) were females. Fisher suggested it represented a generational shift and a new contention led by empowered young women and girls. Cross-cutting representation makes this group a powerful messenger for a disruptive narrative.

One clear sign that the youth movement has been effective with its message: the teen girls are getting attacked. Hateful messages from climate deniers, misogynists, and far-right agitators[85] berate the teen girls' appearance, race, age, and ideas. Thunberg has borne the brunt of the hate speech (including via tweet by President Trump) but teen girl activists (and their families) around the globe have received similar verbal threats. The sexism ingrained in these attacks is sadly not unique; experts note that a pervasive culture of mockery and abuse is present across online cultures, often manifested in attacks on women and girls.[86] This may be a reaction to the activists' call for social change and upheaval to avoid ecological collapse; opponents try to solidify their own identities and beliefs by attacking those calling for change. *Columbia Journalism Review* stated that the striking teenagers have become "the stars of an optimistic new climate-change media narrative."[87]

The youth narrative is important for an additional reason: the pass-through effect of their views on their parents' beliefs.[88]

In 2019, Thunberg presented compelling testimony at the World Economic Forum in Davos, Switzerland,[89] was nominated for a Nobel Peace Prize, and named Person of the Year by *Time* magazine. During the COP 25 climate summit in Madrid, Thunberg and dozens of young climate activists from around the world took to the stage and chanted:

We are unstoppable! Another world is possible!

CULTURAL TRANSFORMATION

This chapter took a broad view of fossil fuel culture—its history, its consumerism, and how this deep culture might be transformed. Appropriately, its transformation will be a difficult and multifaceted challenge for all levels of society. To be honest, I sometimes worry about our collective ability to change fossil fuel culture dramatically enough—and quickly enough—to preserve the healthy functioning of Eairth. But at other times, I feel quite optimistic because history demonstrates that cultures have changed previously, and sometimes rapidly.

Scholars have applied the analogy of the abolition of slavery to climate change (of course, recognizing their *very* significant differences) as an example of how culture can change in radical ways.[90] Social order responses to abolitionists' demands strike some familiar notes: *Slave labor is our primary source of "energy" and we must continue it. We cannot change the system and life as we know it. The system is too large to dismantle and the economic impacts too large. Political resistance is too great and vested interests are strongly opposed.* Yet, after decades of struggle (and violence), slavery was abolished. Andrew Hoffman, professor of business sustainability and organization, said the analogy gives us reason to be hopeful: we have faced similar challenges and overcame them. In the end, hope is the critical element. "Through the arc of history, hope and belief against the odds accomplished far more in motivating people to action than did data and models."[91] Another world and culture is indeed possible.

Understanding that fossil fuel culture must change is a vital precursor to understanding how we as individuals can best contribute to that change. After all, cultures consist of individuals, some of whom may take to the streets but far more who will not. If we accept the 3.5 percent rule of mass civil disobedience, what's the role for the remaining 96.5 percent in moving toward a carbon-free future? They are vital in so many other ways. Because cultural transformation involves all levels of society, individuals like you are the "insiders" in those places. You are inside a family and circle of friends, a workplace, a club, a voting booth (I hope), a church, a neighborhood. You are a purchaser of material culture and a target of symbolic culture. You don't have to know how to "solve" climate change, but you recognize that social change is needed urgently to address the climate crisis. Like all of us, you are currently embedded in a fossil fuel culture, and you play an important role in ushering in a new culture that will help you and Eairth flourish.

NOTES

1. Judith Shulevitz, "Alexa, Should We Trust You?," *The Atlantic*, November 2018, https://www.theatlantic.com/magazine/archive/2018/11/alexa-how-will-you-change-us/570844/.
2. Adam Greenfield, *Radical Technologies: The Design of Everyday Life* (London and New York, NY: Verso, 2018).
3. Aileen Nowlan, "E-Commerce's Sustainability Problem Isn't Just the Packaging," *Green Biz*, July 23, 2019, https://www.greenbiz.com/article/e-commerces-sustainability-problem-isnt-just-packaging.
4. Stephanie LeMenager, *Living Oil: Petroleum Culture in the American Century* (New York, NY: Oxford University Press, 2014).
5. Amitav Ghosh, *The Great Derangement: Climate Change and the Unthinkable* (Chicago, IL: University of Chicago Press, 2016).
6. LeMenager, *Living Oil*, 11.
7. Barry Lord, *Art & Energy: How Culture Changes* (Washington, DC: AAM Press, 2014); Frederick Buell, "A Short History of Oil Cultures: Or, the Marriage of Catastrophe and Exuberance," *Journal of American Studies* 46 (2012): 273–93.
8. Charles C. Mann, "Peak Oil Fantasy: Why Counting on Scarcity Is Courting Disaster," *Orion*, September–October 2015, p. 16.
9. Quoted in Mann, "Peak Oil Fantasy," 16.
10. Mann, "Peak Oil Fantasy," 16.
11. Buell, "A Short History of Oil Cultures," 277.
12. Buell, "A Short History of Oil Cultures," 290.
13. LeMenager, *Living Oil*.
14. Mann, "Peak Oil Fantasy," 17.
15. K. Levin, B. Cashore, S. Bernstein, and G. Auld, "Overcoming the Tragedy of Super Wicked Problems: Constraining Our Future Selves to Ameliorate Global Climate Change," *Policy Sciences* 45, no. 2 (2012): 123–52.
16. Buell, "A Short History of Oil Cultures."
17. Buell, "A Short History of Oil Cultures," 291.
18. Buell, "A Short History of Oil Cultures," 284.
19. Buell, "A Short History of Oil Cultures," 274.
20. Ghosh, *The Great Derangement*.
21. Lord, *Art & Energy*, viii.
22. Lord, *Art & Energy*, ix.

23. Buell, "A Short History of Oil Cultures," 275.
24. Buell, "A Short History of Oil Cultures."
25. George Ritzer, *Enchanting a Disenchanted World: Revolutionizing the Means of Consumption* (Thousand Oaks, CA: Pine Forge Press, 2010).
26. Ritzer, *Enchanting a Disenchanted World*.
27. Thomas Hine, *I Want That!: How We All Became Shoppers*, 1st edition (New York, NY: HarperCollins Publishers, 2002).
28. Julia Corbett, *Out of the Woods: Seeing Nature in the Everyday*, 1st edition (Reno, NV: University of Nevada Press, 2018).
29. Julia B. Corbett, *Communicating Nature; How We Create and Understand Environmental Messages* (Washington, DC: Island Press, 2006).
30. Fiona Brannigan, "Dismantling the Consumption-Happiness Myth: A Neuropsychological Perspective on the Mechanisms That Lock Us in to Unsustainable Consumption," in *Engaging the Public with Climate Change: Behaviour Change and Communication*, ed. Lorraine Whitmarsh, Saffron O'Neill, and Irene Lorenzoni (New York, NY: Routledge, 2015), 84–99.
31. Brannigan, "Dismantling the Consumption-Happiness Myth."
32. Carmella De Los Angeles Guiol, "In Real Life: The Question of Connection in the Digital World," *Orion*, Summer 2018.
33. Janis Dickinson, "The People Paradox: Self-Esteem Striving, Immortality Ideologies, and Human Response to Climate Change," *Ecology and Society* 14, no. 1 (2009).
34. Brannigan, "Dismantling the Consumption-Happiness Myth," 92.
35. Emily Atkin, "You Will Have to Make Sacrifices to Save the Planet," *The New Republic*, June 3, 2019, https://newrepublic.com/article/154036/will-make-sacrifices-save-planet.
36. Rosemary Randall, "Loss and Climate Change: The Cost of Parallel Narratives," *Ecopsychology* 1, no. 3 (2009), doi:10.1089/eco.2009.0034.
37. See www.climatenetwork.org.
38. "Report on Effects of a Changing Climate to the Department of Defense."
39. Jared Diamond, *Collapse: How Societies Choose to Fail or Succeed* (New York, NY: Penguin Books, 2014).
40. Niccolo Machiavelli, *The Prince* (New York, NY: Oxford University Press, 1532 [1952]), 55.
41. Robert J. Brulle and Kari Marie Norgaard, "Avoiding Cultural Trauma: Climate Change and Social Inertia," *Environmental Politics*, 2019, p. 11.
42. Brulle and Norgaard, "Avoiding Cultural Trauma."
43. John Cook, Geoffrey Supran, Stephan Lewandowsky, Naomi Oreskes, and Edward Maibach, *America Misled: How the Fossil Fuel Industry Deliberately Misled Americans about Climate Change* (Fairfax, VA: George Mason University Center for Climate Change Communication, October 14, 2019), https://www.climatechange-communicat ion.org/america-misled/.
44. Jonathan Watts, Jillian Ambrose, and Adam Vaughan, "Oil Firms to Pour Extra 7m Barrels per Day into Markets, Data Shows," *The Guardian*, October 10, 2019, https://www.theguardian.com/environment/2019/oct/10/oil-firms-barrels-markets.

45. Robert J. Brulle, Melissa Aronczyk, and Jason Carmichael, "Corporate Promotion and Climate Change: An Analysis of Key Variables Affecting Advertising Spending by Major Oil Corporations, 1986–2015," *Climatic Change*, December 11, 2019, doi:10.1007/s10584-019-02582-8.

46. Brulle and Norgaard, "Avoiding Cultural Trauma."

47. Jeffrey C. Alexander, *Trauma: A Social Theory* (Cambridge: Polity, 2013); P. Sztompka, "The Trauma of Social Change," in *Cultural Trauma and Collective Identity*, ed. Jeffrey C. Alexander (Berkeley, CA: University of California Press, 2004), 155–95.

48. Ron Eyerman, *Is This America? Katrina as Cultural Trauma* (Austin, TX: University of Texas Press, 2015).

49. Sztompka, "The Trauma of Social Change," 66.

50. Brulle and Norgaard, "Avoiding Cultural Trauma."

51. Sztompka, "The Trauma of Social Change."

52. Brulle and Norgaard, "Avoiding Cultural Trauma," 18.

53. Brulle and Norgaard, "Avoiding Cultural Trauma," 15.

54. Alexander, *Trauma: A Social Theory*.

55. Emily Nicolosi and Giuseppe Feola, "Transition in Place: Dynamics, Possibilities, and Constraints," *Geoforum* 76 (2016): 153–63.

56. Amanda Smith, "The Transition Town Network: A Review of Current Evolutions and Renaissance," *Social Movement Studies* 10, no. 1 (2011): 99–105.

57. https://transitionnetwork.org/.

58. Philip Smith and Nicolas Howe, *Climate Change as Social Drama: Global Warming in the Public Sphere* (Cambridge: Cambridge University Press, 2015).

59. Alexander, *Trauma: A Social Theory*.

60. J. Gosling and P. Case, "Social Dreaming and Ecocentric Ethics: Sources of Non-Rational Insight in the Face of Climate Change Catastrophe," *Organization (London)* 20, no. 5 (2013): 705–21, 708.

61. James W. Carey, *Communication as Culture: Essays on Media and Society* (New York, NY: Routledge, 2009).

62. Robert D. Bullard, *Dumping in Dixie: Race, Class, and Environmental Quality* (Boulder, CO: Westview, 2000).

63. Brulle and Norgaard, "Avoiding Cultural Trauma," 6.

64. Robert J. Brulle, "From Environmental Campaigns to Advancing the Public Dialog: Environmental Communication for Civic Engagement," *Environmental Communication* 4, no. 1 (2010): 82–98, 88.

65. Julia B. Corbett, "Media, Bureaucracy, and the Success of Social Protest: Newspaper Coverage of Environmental Movement Groups," *Mass Communication & Society* 1, no. 1 and 2 (1998): 41–61.

66. Corbett, "Media, Bureaucracy, and the Success of Social Protest."

67. Extinction Rebellion Staff, *This Is Not a Drill: An Extinction Rebellion Handbook* (London: Penguin Books, 2019).

68. John Lyons, "A Wave of Protest Rattles Governments – Around the Globe, Messaging Apps Spur Mass Movements against the Status Quo," *Wall Street Journal*, November 23, 2019.

69. Jan Selby, Omar S. Dahi, Christiane Fröhlich, and Mike Hulme, "Climate Change and the Syrian Civil War Revisited," *Political Geography* 60 (September 1, 2017): 232–44, doi:10.1016/j.polgeo.2017.05.007.

70. A. Leiserowitz, E. Maibach, S. Rosenthal, J. Kotcher, and P. Bergquist, *Climate Activism: Beliefs, Attitudes, and Behaviors, November 2019* (New Haven, CT: Yale University and George Mason University, 2019).

71. Erica Chenoweth and Maria J. Stephan, *Why Civil Resistance Works: The Strategic Logic of Nonviolent Conflict* (New York, NY, USA: Columbia University Press, 2011).

72. David Robson, "The '3.5% Rule': How a Small Minority Can Change the World," *BBC News*, May 13, 2019, https://www.bbc.com/future/article/20190513-it-only-takes-35-of-people-to-change-the-world.

73. Chenoweth and Stephan, *Why Civil Resistance Works*.

74. Robson, "The '3.5% Rule'."

75. Extinction Rebellion Staff, *This Is Not a Drill*, 100.

76. Sam Knight, "Does Extinction Rebellion Have the Solution to the Climate Crisis?," *The New Yorker*, July 21, 2019, https://www.newyorker.com/news/letter-from-the-uk/does-extinction-rebellion-have-the-solution-to-the-climate-crisis.

77. Extinction Rebellion Staff, *This Is Not a Drill*.

78. Aída Chávez, "Climate Activists from Extinction Rebellion Glued Themselves to the Capitol to Disrupt House Votes," *The Intercept*, July 24, 2019, https://theintercept.com/2019/07/23/extinction-rebellion-climate-change-capitol-hill.

79. Hayes Brown, "These Pictures Of Activists Begging You To Take Climate Change Seriously Are Intense," *Buzz Feed News*, July 12, 2019, https://www.buzzfeednews.com/article/hayesbrown/15-pictures-extinction-rebellion-climate-protests-europe.

80. Chloe Farand, "'Climate Emergency': How Extinction Rebellion's Language of Urgency Is Shaping the Political Agenda," *Desmog UK*, May 1, 2019, https://www.desmog.co.uk/2019/05/01/climate-emergency-extinction-rebellion-language-urgency-parliament-labour-motion-ccc-net-zero.

81. Homa Khaleeli, "'I Would Go to Prison': The Ordinary People Getting Arrested for Extinction Rebellion," *The Guardian*, April 23, 2019, https://www.theguardian.com/environment/2019/apr/23/i-would-go-to-prison-the-ordinary-people-getting-arrested-for-extinction-rebellion.

82. Knight, "Does Extinction Rebellion Have the Solution."

83. Greta Thunberg, "I'm Striking from School to Protest Inaction on Climate Change – You Should Too," *The Guardian*, November 26, 2018, https://www.theguardian.com/commentisfree/2018/nov/26/im-striking-from-school-for-climate-change-too-save-the-world-australians-students-should-too.

84. Sara Kaplan, "Teen Girls Are Leading the Climate Strikes and Helping Change the Face of Environmentalism," *Washington Post*, September 24, 2019, https://www.washingtonpost.com/science/2019/09/24/teen-girls-are-leading-climate-strikes-helping-change-face-environmentalism/.

85. Linda Givetash, "Young Female Climate Activists Face Hateful Abuse Online," *NBC News*, November 10, 2019, https://www.nbcnews.com/news/world/young-female-climate-activists-face-hateful-abuse-online-how-they-n1079376.

86. Givetash, "Young Female Climate Activists."

87. Abby Rabinowitz, "How Teen Climate Activists Get—and Make—Climate News," *Columbia Journalism Review*, May 1, 2019, https://www.cjr.org/united_states_project/climate-strike-high-school-teenagers.php.

88. Lydia Denworth, "Children Change Their Parents' Minds about Climate Change," *Scientific American*, May 6, 2019, https://www.scientificamerican.com/article/children-change-their-parents-minds-about-climate-change/.

89. Greta Thurnberg, "'Our House Is on Fire': Greta Thunberg, 16, Urges Leaders to Act on Climate," *The Guardian*, January 25, 2019, https://www.theguardian.com/environment/2019/jan/25/our-house-is-on-fire-greta-thunberg16-urges-leaders-to-act-on-climate/.

90. Andrew J. Hoffman, *How Culture Shapes the Climate Change Debate* (Stanford, CA: Stanford University Press, 2015); W. Nuttall, "Slaves to Oil," *EPRG Working Paper No. 0921*, University of Cambridge Electric Policy Research Group, 2009.

91. Hoffman, *How Culture Shapes the Climate Change Debate*, 79.

Chapter 3

Individuals as Social Actors, Not Consumers

In the 1970s after the first Earth Day, I read lots of "lists," like Ten Things to Do to Green the Earth, or Fifty Simple Ways to Save the Planet. The lists instructed me to dry clothes on a clothesline, turn off my lights, have a vegetarian day, turn down my thermostat, plant a garden, and shower with a friend to save water (hey, it was the 1970s).

Well, those lists are back. Many of the personal steps are exactly the same as fifty years ago, with some technological updates: pay bills online, turn off computer overnight, work remotely from home, and switch to LED lights. Now the lists are called 20 or 101 Things to Do to Stop Global Warming.

These small, individual actions are just as laudable today as they were fifty years ago, and using less energy at home and living more lightly on the planet is a no-brainer. I take these little steps because they make me feel better, and because fossil fuel culture has taught me to take actions in accordance with my consumer identity. But are such personal steps able to "stop global warming"? Sadly, no. Even if everyone in my city, state, and nation undertook *all* 101 actions (which they won't—or can't), the reductions in greenhouse gas emission would be wholly insufficient to address the climate crisis. It's a matter of scale; in a world largely run by fossil fuel energy, large-scale, systemic change is necessary.

If you feel a bit deflated by that news, think of it instead as a positive and important redirection of your time and energy. I hear enormous guilt and shame from individuals who blame themselves for not being "green enough," for needing to drive to work, for not being able to afford solar panels, for getting plastic containers from the deli. When voluntary, consumer-oriented personal action is framed as the "solution," then climate change seems like your personal fault. It's not. The laser focus on consumer actions is a bit of a foil; it distracts us from the large-scale transformation of fossil fuel culture that is the real solution—and from the vital role that individuals need to play in that transformation.

Much of current climate communication persuades individuals to take voluntary, consumer actions. By all means, keep doing the environmentally friendly things, but think of them instead as a *bridge* to communicating and taking action in much more significant ways—ways that influence the social, cultural, and structural levels. Even if you don't want to join Extinction Rebellion, your involvement as a socially engaged individual—with strong ties and shared values with collectives such as family, friends, workplace, church, and social groups—is very important. The new direction presented by this chapter demonstrates that you have far more power to help address the climate crisis as a social actor than as a consumer.

RETHINKING THE ROLE OF THE INDIVIDUAL

When cold weather returns to Salt Lake City, where I live, so do the air inversions: a high-pressure weather system parks on top of the valley between two mountain ranges, and the "lid" traps all the pollution we produce daily, which triggers health warnings. My throat is scratchy, my sinuses unhappy, and breathing is uncomfortable. Half of the gunk is produced by "mobile sources" like cars and trucks, and the rest comes from industry and power plants (which produce electricity). Only a strong storm will blow the pollution away (well, blow it somewhere else).

The messages I hear during an inversion are all about actions individuals should take. The news warns the young, the elderly, and "sensitive groups" to stay indoors, and asks us to drive less or carpool. Kids must stay inside for recess. City and state officials ask individuals to take mass transit. A brief run of TV ads asked residents to turn down the thermostat and combine their errands. None of the air inversion messages connects the pollution to climate change, although the fossil fuel combustion trapped as air pollution is the same carbon emissions that thicken the atmospheric blanket that warms the planet.

In fossil fuel-based culture, you have been well-schooled (as those air inversion messages show) to think that your most effective role as an individual is as a consumer taking small, voluntary actions. Recycle. Become a vegan. Bring reusable bags. Eschew air travel. So just how effective are individual actions in reducing carbon emissions to slow climate change?

A Matter of Scale

When you drive by the refinery on north side of Salt Lake City, you can see yellow flames coming from several smokestacks. That is "flaring," when gas (that is considered unwanted or excess) is burned off at refineries, chemical plants, landfills, and during oil and gas extraction. Although capturing the gas is an alternative to flaring, it is not often done. Gas flaring contributes to air pollution and climate change and has a substantial impact on the health of those living close by. The World Bank estimated that worldwide in 2018, the amount of gas flared was about as much used in Central and South America in a year.[1] Another study estimated that flaring produces 281 million tons of CO_2 annually, or the equivalent of what's emitted by 60 million passenger vehicles driven for one year, or 32.5 million homes' energy use for one year.[2] It puts into perspective a message telling you to drive less.

Another study considered a broad range of individual actions and calculated their ability to reduce greenhouse gases. The conclusion was that commonly promoted things like recycling and changing light bulbs don't move the needle.[3] By far, the action that substantially reduced personal emissions was having one fewer child, which in a developed country amounts to 58.6 tons of CO_2e (CO_2-equivalent) saved annually. The next three actions pale in comparison: living car-free (just 2.4 tons CO_2e saved per year), avoiding airplane travel (1.6 tons CO_2e saved per roundtrip transatlantic flight), and eating a plant-based diet (0.8 tons CO_2e saved per year). Though the authors noted that individual actions can sometimes create ripple effects and shift socially acceptable behavior, the fact remains that only fractions of the global population would ever undertake these behaviors voluntarily. To slow global warming, huge reductions needed are far beyond what individuals can do in their homes and daily lives.[4]

Personal Change versus Social Change

These numbers tell us that it's important not to conflate *personal change* with *social change*; voluntary personal behaviors and a "green" lifestyle—admirable though they are—simply do not come close to the scale of social and cultural change required.[5]

Some critics believe that the overwhelming focus on personal consumer action is undermining the global fight against climate change. One study found that pointing fingers at individual lifestyles distracted from large-scale solutions, such as getting big corporations to ditch fossil fuels and governments to adopt policies like carbon taxes.[6] A business professor concluded that focusing on small consumer actions can "trivialize the challenge of global warming, and divert our attention from the huge technological and policy changes that are needed to combat it. . . . It is absurd for middle-class citizens in advanced economies to tell themselves that eating less steak or commuting in a [Prius] will rein in rising temperatures. To tackle global warming, we must make collective changes on an unprecedented scale."[7]

So, does that mean we should put on shorts and crank up the heat? No. We should all live more respectfully and lightly on the planet. As climate scientist Michael Mann wrote, "[These actions] save us money, make us happier and healthier, and set a good example to follow. But these actions are no substitute for systemic solutions that incentivize a collective shift" to reducing our reliance on fossil fuels for energy and transport, which accounts for about two-thirds of global carbon emissions.[8] Exactly. The point is not to be misled that reducing your personal energy use is where you *stop*. Instead, think of consumer choices as a *bridge* to motivate you (and others) to participate in system-wide social changes that are effective.

There is so much attention to personal actions to "stop global warming" that it has entered the fray of a culture war. When the US Green New Deal (which proposes massive investment in renewable energy jobs) was announced, Republicans held a press conference to attack it, saying it would "outlaw" burgers (and a congressman ate one in front of the cameras). Even among climate activists, particular actions have become a kind of purity test —"virtue signaling"[9] how serious a person must really be about the issue. "Flight shaming" is now a thing.

Blaming ourselves for not tweaking our consumptive habits far enough is misdirected and I would say dangerous. An opinion piece in the *New York Times* said,

> It turns eco-saints against eco-sinners, who are really just fellow victims. It misleads us into thinking that we have agency only by dint of our consumption habits – that buying correctly is the only way we can fight climate change. As long as we are competing for the title of "greener than thou," or are paralyzed by shame, we aren't fighting the powerful companies and governments that are the real problem. And that's exactly the way they like it.[10]

Instead, redirect your focus and your energy from yourself to a larger collective—whether that's your soccer club, workplace, or your city and its air pollution problem.

Additional Factors and Constraints

An important part of rethinking the role of individuals is recognizing the tension in asking someone to take personal action when that action will only be successful at a widespread collective level—and even then, very real barriers and challenges to participation remain.

For example, how do individuals react to messages about an air inversion? Of course, some people will take small steps like combining errands. However, a great many folks do not talk about or acknowledge the bad air;

some continue exercising outdoors (despite warnings). Bus ridership does not increase. A few people wear medical face masks (which do not filter out lung-damaging particles). Others cope by escaping to the ski resorts above the gunk. Obviously, the messages directed toward voluntary individual action have not been successful in reducing air pollution. One reason is that individual drivers produce only half of the emissions. Also, many more factors go into decision-making than a simple message or two, and people face a wide range of barriers—practical, psychological, and physical—to taking action.

Climate change (and all environmental issues) are classic *collective action problems*,[11] where sufficient collective action benefits everyone, though some individuals will not comply, thinking the action too "costly," not in their self-interest, or physically impossible. For example, if I take the bus to work, I helped protect the air we hold in common and reduced greenhouse gas emissions, even though I received no immediate, tangible benefit for my action (other than my conscience). Meanwhile, my neighbor receives no immediate harm for his driving, even though he benefits from those (like me) who didn't drive—the classic "free rider" effect. And, my neighbor's free-riding may be entirely beyond his personal control—the result of the car-based architecture of my city and lack of mass transit options.

The collective action problem of climate change cannot be addressed by small voluntary actions because the problem was created and is maintained by a collective—all of us. And, it involves the collective "resource" of our planetary climate system. Solving a collective action problem requires the collective participation of *all* people, which could be secured through regulations, fines and enforcement, and new technology, or changes in infrastructure, the marketplace, and the physical environment. (Just in the transportation sector, imagine what that might look like.) Communication is needed far beyond the individual level to move the collective to action that benefits all.

An example of required participation for a collection action problem is China's plan to reduce single-use plastics across the country: no single-use straws in restaurants and no plastic bags in major cities by the end of 2020.[12] Restaurant and hotel industries must greatly reduce or eliminate all single-use plastic items. Since China produces even more plastic waste than the US, required collective participation by businesses and individuals is a significant first step.

At an individual level, decisions regarding your fossil fuel consumption are often not simple choices made by an autonomous person. You face numerous constraints, be they economic, ideological, physical, or normative. *Social norms* (how your social peers use energy or how they believe it should be used or valued) influence your energy consumption, but not always in a positive direction. One campaign to reduce home energy use[13] discovered that when people learned their neighbors used more energy than they did, their own usage increased. What affects your individual decision to use (or not use) mass transit? Chances are, it's family and peers' social norms about mass transit, as well as the physical environment. A friend of mine lives close to a bus line but she has never ridden the bus; she lacks normative social support because bus-riding just isn't something her friends, family, or coworkers ever do. Culturally we are trained to be drivers, not bus-riders, and the physical environment in many cities is stacked high in favor of cars. There is much that could be done *above* the level of the individual to make driving less convenient and desirable and mass transit more so.

Similar constraints affect many decisions about your energy use. Even if you would love an electric car[14] or a more efficient furnace, only a privileged few can afford them (an economic constraint). Otherwise, energy is pretty cheap and affordable, so people leave doors and windows open (in cold or hot weather) and this seems acceptable (a social norm influence). When you buy a car, you are constrained by your budget and by the types of autos manufactured (an industry-level constraint) and what kind of gas mileage the car gets (primarily a government standard). One-time-use disposable products are so cheap that the energy used to create (and dispose of) them never crosses your mind (both economic and marketing influences). Perhaps the affordable apartment you rent is near an oil refinery; people with less social and economic power face greater environmental harm, an example of environmental injustice. Placing the burden of action solely on the individual for reducing fossil fuel use ignores significant additional constraints not under individual control: the utility's energy source, state building standards for insulation and construction, community walkability, and electronic appliance design and pricing.

Individuals remain vital players in the climate crisis. But for a variety of reasons, the messages about voluntary actions you receive contribute more often to the social inertia around climate change (and feelings of guilt and shame), rather than working toward social change. We need to rethink the best role for individuals to make a difference, which includes rethinking the traditional ways that communicators try to change attitudes and consumer behaviors.

INFORMATION AND THE INDIVIDUAL

In the US, a person is exposed to roughly 5,000 advertisements each day. Even though you pay attention to very few of those, nevertheless these ads implore you to buy during all your waking hours (continually reinforcing your consumer identity). In comparison, you see extremely few messages that don't ask you to buy, or that try and influence your attitude or behavior in a pro-climate kind of way. So, it's important for those messages to make a mark.

When you see a message about climate change—on TV news, via social media, in conversation—you bring a lot to the table when you evaluate it: what you think and know about climate change, fear or anxiety, and perhaps attachment to a place where you've witnessed change. Given all that, what kind of message might you pay attention to?

That's what climate change communication researchers try to figure out. They test how reception to messages differs based on individuals' personal characteristics. Characteristics might include age or sex, political or environmental ideology (a system of beliefs), income, awareness and understanding of climate change, and *self-efficacy* (beliefs about whether you believe you can help address climate change, or have the skills or knowledge to know how to). One message might require deep thought processing, another message could invoke a fear appeal or an appeal to moral responsibility, and one might frame climate change as high risk and urgent. Other variables might test whether it makes a difference what kinds of news or social media you pay attention to. The logic of this "matching" is that individual characteristics may somewhat predict the type of message a person might respond to—for example, women who are politically conservative, live in the Midwest, and are "doubtful" about climate change might respond to an economic message. Or, perhaps liberal California men would respond to an altruistic appeal.

There are many good resources that summarize this research and provide tips to communicators.[15] Generally, the advice is to avoid overwhelming someone with science, facts, or graphs. Instead, connect with a person's values and worldviews; focus messages on solutions and benefits; use messengers who are trusted and credible; use images, stories, and mental models; and frame your message around issues your audience cares about, whether health or national security.

That's all good advice. Yet, the underlying assumption in this research is that *information* is the problem—that we just need to connect the right message with the right person, and then people will change their attitudes and behaviors and climate change will be addressed. Even some scientific reports (and some scientists) stand by the "more information, more knowledge" solution. When the Fifth Assessment Report of the IPCC (Intergovernmental Panel on Climate Change) was released, Chair Rajendra Pachauri said, "We have the means to limit climate change. . . . All we need is the will to change, which we trust will be motivated by knowledge and an understanding of the science of climate change."[16]

Even though research has largely dispelled this "information deficit" model of communication,[17] it remains the predominant type of climate communication research and practice. However, typical one-way communication (both individual messages and entire campaigns) rarely create deep engagement and have not led to sustained behavioral or political participation.[18] More or "better" information and knowledge alone are simply not powerful enough to engage the public and produce action. There is even evidence that more information can make things worse, hindering behavior change,[19] and decreasing concern and a sense of personal responsibility.[20] One study found that people stopped paying attention to climate change when they realized there was no easy solution to it.[21]

Even highly personalized information might not produce action. One study found that calculating a personal "carbon footprint" didn't result in someone using less energy.[22] An individual faces barriers when he tries to reduce his energy use such as the physical environment (your city lacks a good mass transit system), important social factors in your peer groups, and psychological issues such as the helplessness people feel in the face of climate change.

And, message design elements pale when compared to the role of political ideology. In a large analysis across fifty-six nations regarding climate change beliefs, the single strongest correlation was political affiliation and political ideology.[23] How much a person knew about climate change didn't matter. Partisan beliefs increasingly polarize us; conservative white males are more likely than all other groups to support climate denial, and this group had a significant influence on public understanding.[24]

This is not to say that no more scientific information or policy options need be communicated. There remains significant confusion and conflation of climate change with ozone depletion, air pollution, and weather.[25] And "belief" in climate change continues to wax and wane with heat waves and blizzards, as does knowledge of the overwhelming scientific consensus. Individuals also lack practical information about how to drastically reduce emissions.

Although we often think of individuals as largely independent in their actions, we are profoundly shaped by social, political, and economic spheres that can render the strongest of our behavioral intentions impotent. Information is important at all levels of culture, but in far broader and different ways than messages targeting individual, consumer-oriented actions. Individuals do indeed have an important role to play regarding climate change, but it's different from what you might have imagined. According to an editorial by researchers at the Center for Climate Change Communication at George Mason University, that role is collective action:

> While individual behavior changes can reduce emissions, their contributions are insufficient in the absence of large-scale, systemic change. For emissions to rapidly fall, the policies, regulations, and technologies that shape our energy use must change in ways that promote sustainable lifestyles and remove existing barriers to sustainable action. These changes are more likely to be made if

citizens and consumers demand them. Thus, collective action by citizens and consumers is sorely needed to prod legislators and corporations into enacting the policies and practices that can stabilize the climate.[26]

INDIVIDUALS ARE SOCIAL ACTORS

Let's start with a simple example: why are you wearing what you're wearing today? I'm wearing black yoga pants and a cotton-poly sweater. Although I made an individual choice to purchase and wear this, my choice was greatly influenced by what my social peers wear, what the fashion industry makes available, and by global economic factors that determined the price and influenced where materials were obtained, the wage workers who assembled them, and global shippers that delivered them. Suddenly, my bedroom seems rather crowded.

For something so seemingly simple, your thoughts and actions regarding your clothes are more a product of complex social, cultural, and economic interactions than they are a simple, autonomous choice you make each morning. As a highly "social animal," you utilize the thinking and feeling parts of your brain simultaneously and continuously to evaluate messages, cues, and emotions from others. These influences help form your identity and a sense of belonging. Your brain relies on powerful social norms and social pressures to shape your sense of what's acceptable and validated both in peer groups and larger cultural settings. Bottom line: we learn through and are products of social interaction.

To help you start conceptualizing yourself as embedded in a social and cultural system rather than as a lone individual, think about each action or decision you made in the last several hours. What or who had an influence on what you ate, how you traveled, media you read, or what you talked about? From where do *your* influences and interactions most often come?

Social Actors in Social Groups

Deciding what you pay attention to or ignore (and what you do or don't do) is a *social* process you learn in interaction with the people and environment around you.[27] In other words, your assessment of the world and your actions in it are *socially constructed*.[28] When you attune to what social others tell you is noteworthy (or of no value), you are a *social actor*, not a wholly independent individual. It's how you learned that plastic straws are "bad for the environment" but that idling is okay (although both concern fossil fuels).

Individuals are strongly influenced by their social groups: family of origin; peers and friends; fellow employees; members of faith, athletic, social, and recreational groups; and so on. Some scholars maintain that our belief systems are formed to a pretty large extent by the groups to which we belong.[29] You generally endorse positions that most directly reinforce your connection with others in your social groups, which also strengthens your identity. If those in your groups commute via bus or bike, chances are that you will, too.

Some social media "friends" may be socially distant from you, but they represent a cultural community and provide important cues as to what is relevant day to day. These social contacts also affect a person's preferred information sources (which we'll return to shortly)—whether newspapers or Facebook, Fox News or public television. However, the ties you have with social media "friends" are generally considered "weak ties," not the strong, personal ties you have with members of a close-knit social group.

Because you are so influenced by and bonded to your social groups, it follows that you have more power to do something about the climate crisis with these important social others than by yourself. Some scholars consider these strong ties a requirement for social change.[30]

Powerful Social Norms

Perhaps you heard recently that you shouldn't use plastic straws because they were bad for the environment. Maybe you heard it from a friend or through a social media post, or you saw that disturbing picture of a plastic straw through the nasal cavity of a sea turtle. Maybe you bought and now carry a reusable straw or don't use one at all; maybe you've been "shamed" for using one. That's the power of social norms.

A *social norm* is a "rule" of conduct that influences an individual's behavior because important social others and peers have made clear what is considered socially acceptable behavior.[31] Social norms are "enforced" by perceived sanctions or ostracization, or by rewards, from your important social others.[32] You don't litter if you don't want your social others to think you are the kind of person who would do that. Social norms are powerful; they influence our attention, conversations, and actions. They may be guided by fairly invisible social forces, but they can shape what we think and feel, talk about, and do. Not surprisingly, people tend to underestimate the influence of social norms on their own behaviors.[33]

Social norms are categorized either as a *descriptive social norm* (a belief about what others are presently doing) or an *injunctive social norm* (a belief about what others should be doing or what most people approve of others doing).[34] Idling a vehicle is an example of a descriptive social norm: when people observe that others frequently idle, it essentially gives them social permission to behave this way. Broadcasting an injunctive norm message telling individuals not to idle is a tough sell if "everyone does it" now.

Social norms play a role in how friends and family perceive climate change. Fewer than half of Americans perceive a social norm in which their friends and family expect them to take action on global warming.[35] Forty-three percent think it is at least moderately important to their family and friends that they take action (an injunctive norm), and four in ten (40 percent) say their family and friends make at least a moderate amount of effort to reduce global warming (a descriptive norm). As a social actor with family and friends, your voice is significant.

Because individuals constantly observe and interact with the social and cultural levels, social norms can appear to change rapidly. Suddenly, an injunctive norm appears that you ought not use a straw, or, you ought to bring your own shopping bag.[36] But norms rarely emerge spontaneously; rather, they are a reflection of underlying material (or consumer) interests[37] and political economy struggles. Say that a city tried to change the social norm of some businesses propping open their front doors (thinking it will attract customers) because it wastes energy, especially in cold and hot weather. The change would contradict larger cultural-level factors: energy is cheap, and energy waste is not connected to anything larger (such as its contribution to air pollution and a changing climate). Even if "ought not prop open doors" emerged as a norm and businesses felt shamed for doing so, the contradictions with the cultural level would remain, and over time the new injunctive norm might be subverted.

Countering Social Group Dynamics

Where did you learn that climate change was important, or, not important? Chances are you received strong cues from one of your social groups – often from your family of origin – as to what to believe. You also are very influenced by your social others about the information sources you use for climate change. While social groups can be good places to become a socially engaged actor and find social support for talking about climate change, it can be hard to challenge existing group social norms ("*Climate change just isn't something we worry about*"), or preferred information sources used to arrive at that conclusion.

Although it might seem that you evaluate a fact solely at the individual level, "facts" too have a communal life, and social attention (or inattention) is very much organized around them.

Candis Callison studied journalists, evangelicals, and socially responsible investors for her book, *How Climate Change Comes to Matter: The Communal Life of Facts*.[38] She found that facts have a rich communal life, gaining meaning and relevance through social interaction. In other words, a "fact" isn't simply a "fact" but something interpreted and contextualized by social others. And, these social others also influence which information sources you attend to.

Confirmation bias is when you give greater weight to (and seek out) information that supports your preexisting beliefs.[39] The filters employed by your belief system are powerfully shaped by social group identification, so much so that "facts" become less important than ideological affiliation for polarized issues,[40] such as climate change.[41] Confirmation bias is often measured as an individual attribute although its roots are socially derived. If your family of origin or other social groups pay attention to climate-change-denying information, this may be the information you have relied on as well.

Another important group dynamic is that your "belief" in climate change wasn't independently discovered one day all by yourself; instead, it was socially informed and maintained. In her ethnography about a small Norwegian town during a very warm and snowless winter, sociologist Kari Norgaard found that residents ignored climate change in response to social circumstances and carried out through social interactions, resulting in *socially organized denial*.[42] Individual Norwegians *collectively* distanced themselves from the climate reality right in front of them (lack of snow and warm temps) because of norms of emotion and conversation (climate change was one of the things "we just don't talk about"). People knew "about it," but social and cultural cues told them it wasn't something to pay attention to (or talk about).

At the psychological level, emotions are the inner feelings that link us to the broader reality—if we pay attention to them.[43] The next chapter discusses denial as a psychological emotional defense, but it's important to talk here about how social groups use social norms to ignore climate change in response to social circumstances and carried out through social interaction: *"Gee, no one in my group is talking about the freakish warm winter without snow, so I won't either."* The social norm is usually to look the other way.

Such active resistance to disturbing (true) information takes a lot of work! Holding something unpleasant at arm's length requires energy, unconscious or not. In this way, Norgaard says that although we "know" about climate change, simultaneously we manage to "not know," which creates a *double reality*. One reality is the pattern of extreme weather events, heat, and rising seas we experience; the other reality is the collectively constructed sense that things remain "normal" in everyday life.[44]

You likely participate in this double reality even if you accept climate science. When you live in the double reality of "knowing" yet "not knowing," you engage in *implicatory denial*, or a failure to integrate the difficult knowledge of the climate crisis into social group interactions and to transform it into social action.[45] In this sense, what is minimized is not the information ("climate change is real" and "this winter is freakishly warm") but the *implications* (psychological, political, or moral) that follow. "Knowledge itself is not at issue, but doing the 'right' thing with the knowledge."[46]

I'd venture that most social groups rarely discuss climate change and its implications for everyday life and practice. The topic now falls beyond what is "normal" to think and talk about. In a social group or society at large, if you perceive that the majority do not hold your belief, there's a good chance you will remain quiet. You don't want to feel excluded or isolated, so you don't speak up. (More about the "spiral of silence" in a later chapter.) And of course, our "perceptions" of what the majority feel or believe is often wrong. For example, most people grossly overestimate the percentage of climate deniers; Australians thought about 22 percent of people were deniers when only about 6 percent were.[47] Perceptions can thus spiral as more social members remain silent.

This is where you come in as a powerful social actor. It is precisely because silence and socially organized denial occur at the social group level that this is the best place to bring forth your feelings and viewpoints. When you are a member of a social group—whether your family of origin or your book club—you have strong relationships that demonstrate care and concern by others in the group. You possess social capital: as a member, you are trusted and valued; you have social agency and can secure benefits and invent solutions to problems.[48] A cohesive group has a shared sense of identity, a shared understanding, shared norms and values, trust, cooperation, and reciprocity. Such a group will listen to your concerns about climate change.

It is key, however, to broach climate change in a way that doesn't appear to discount or ignore the group's established social norms and identity – but instead adheres to them. It's key to talk about your lived experiences and felt emotions and not go down the path of "battling science." Chances are, a silent group member(s) shares your concerns and thoughts. Appealing to shared values (discussed more in a later chapter) can help counter social group dynamics.

Social norms and group dynamics can be changed, sometimes rapidly. It's not a matter of trying to "reprogram" individual behaviors or opinions, but to fully consider the social processes by which individuals internalize why they think and act the way they do about climate change.[49]

HOLISTIC INTERACTIONS LEAD TO SOCIAL CHANGE

As a social actor, you are in constant interaction with three levels: micro, meso, and macro.[50] The micro level involves your individual emotions and actions (which constantly interact with the meso and macro levels). The meso level concerns the social and cultural spheres: social norms of behavior from conversations (e.g., positive support for riding your bike), the feelings of others (family skepticism), and attention (noticing energy waste), and also include accepted cultural habits, customs, and values. Experiences at the meso level shape (and constantly reinvent) what is considered "normal" to think about, talk about, and feel.[51] If climate change is never mentioned in your meso level circles, it likely won't be on your radar either.

The micro and meso are connected to the macro level of political economic relations. That includes the economic system, which is the growth model of industrial capitalism and a marketplace that ships goods around the globe because energy (and labor) is "inexpensive." The macro level also includes physical infrastructure (such as domination of auto transport). The political realm includes laws, regulations (concerning industrial production and pollution), policies (energy costs and subsidies), and international cooperation to address the climate crisis.

Because the micro, meso, and macro levels constantly interact, the causal arrows of influence go both directions, upstream and downstream. Change in one level affects the others. And, if you want successful change at the micro level, it must have congruent support from the meso and macro levels.

Here's an example of how the micro, meso, and macro levels interact. When I walk my neighborhood, I see idling cars, sometimes four or five cars on a cold morning idling in driveways and at curbs. I see cars idling sans drivers outside stores, restaurants, and offices. In front of buildings, people idle cars while waiting to pick someone up or drop something off. People at city parks eat lunch inside their cars with the engines running. People have long conversations in front of their respective idling cars. I watched a woman drop off cardboard boxes at the recycling container while her truck idled. City buses idle at end of their route on campus, right in front of "No Idling" signs. Idling your vehicle is a common, observable behavior—a descriptive social norm of what people are doing.

Salt Lake City took a typical approach: change individuals' behavior with a campaign to reduce idling to help meet air pollution targets (with no mention that emissions contribute to climate change). "Turn the Key, Be Idle-Free" stickers appeared on cars and in public locations, notably schools. Four years later, a no-idling city ordinance passed (though it exempted cars on private property). After some news stories, publicity largely ceased; a decade later, many residents don't know the ordinance exists (in part because it is not enforced).

A typical communication research approach to learn why the idling campaign (and ordinance) did not change behavior would survey individual drivers: did they see or remember idling messages, were they aware of the law, did they idle, and other individual attributes that might correlate with behavior. The results would be used to get "better information" to residents so they understood the problem, which would encourage them to change their behavior. After all, turning the key is a very easy, pro-climate behavior.

What's missing in this approach? An individual's environmental sentiments or actions—about idling, climate change, or nuclear waste—are the product of her social and cultural interactions and influences. A person "learns" to idle by observing others doing it and concludes (logically) that it's acceptable. A person also gets lots of reinforcing messages from the political and economic sectors (it appears legal to idle and gas is cheap). It's easy to see that communication (and action) must extend far beyond pleas to lone individuals. The interactions among these levels have constructed idling as rather irrelevant because "everyone does it." In addition, the largely unknown ordinance is not enforced, and the connection between individual action and the quality of the atmospheric "commons" is not emphasized.

Norgaard explains the contradictions among these interactions (micro, meso, macro) as they relate to climate change.[52] The individual may hold feelings about climate change—concern, powerlessness, and guilt in the context of social pressure to fit in—but might perceive no space for emotional or conversational expression with others (meso). Another contradiction is between present-day behaviors that are antithetical to reduced emissions but are presented as normal in everyday life, such as broad acceptance for wasting or using lots of energy (patio heaters, open doors, and low-mileage vehicles).

In this complex realm, it's not surprising that appeals made to individuals are not particularly effective. The links between social structure and political economy shape individual sentiments and experience, and the individual plays a part in the reproduction of the status quo. A "social fact" (whether idling or the larger car-dependent culture) is "the way in which individuals' seemingly rational actions are in fact merely reflections of permissible patterns of behavior within a particular social structure."[53] For example, it seems rational on any given day for me to use my car (in the social and economic structure in which I live), even though I'm extremely concerned about climate change. Climate communicators (and researchers) are very familiar with the numerous contradictions between stated environmental values and political economy, and knowledge, values, and everyday practice.

If you think of social change *holistically*, behavior change by individuals works best when changes at all levels support and reinforce individual change. Thus, communication (and action) is most successful when taken at *all* levels, not just by individuals. For a collective action problem like climate change, a holistic, comprehensive view of change is most likely to succeed. Here's an example of holistic change that's familiar to many.

In the late 1970s, the newsroom where I worked was a haze of cigarette smoke. People smoked in meetings, conference rooms, lobbies, restaurants, and in the "smoking section" of airplanes. That might sound shocking now, but smoking was the descriptive norm: it's what people did, everywhere. In that era, smoking was hip and glamorous.

So how and why did smokers' behaviors change? Many didn't quit until smoking became highly inconvenient, expensive, a shunned behavior, and illegal in certain locations—conditions created by the larger social and economic systems. After decades of tobacco industry concealment, studies of smoking's health impacts were finally publicized. Soon, laws severely restricted smoking in public places. When secondhand smoke impacts were revealed, society knew that smoking harmed innocent others. Taxes on cigarettes skyrocketed (an economic disincentive). Only after all of this did a broad injunctive norm develop: you should not smoke, nor expose others to it. There were also social events and challenges to stop smoking, like the American Cancer Society's annual smoke-out. The percentage of smokers dropped from 42 percent in the 1960s to 14 percent today.[54]

This example shows that significant social and cultural change was possible, that it required long-term (and continuous) effort, and that the change pressures must be holistic, complementary, and present in all levels: political, economic, in the physical environment, in social groups, and by individuals. Change is less supported (or likely) at the individual level if barriers or inaction exist at other levels. This example makes apparent the weaknesses in the idling campaign, and generally in efforts to curb energy use: the burden for change cannot be placed solely in individuals' laps.

The smoking example also illustrates that change pressures sometimes need to *begin* at the meso or macro level (such as regulation, taxation, and health warnings) and filter down to individuals through interactions and constraints. In other cases, change efforts might begin at one level, or simultaneously among levels. This has been the case in various US cities working to increase bicycle use. Addressing physical barriers (lack of dedicated bike lanes, bike parking, bike rentals, and safety) has been crucial in cultivating a "bike culture" that motivated and better protected riders. Another macro-level effort was when a thousand cities and jurisdictions set zero-emission goals and declared "climate emergencies" to apply social pressure on all levels and further political action.[55]

Another example of change that began at the meso and macro levels involves pesticides. Because there is tremendous normative pressure to use lawn chemicals and achieve "perfect" lawns, it's unlikely that messages to individuals to stop using chemicals would be effective. In Canada, homeowner associations and neighborhoods (meso level) came together to ban pesticide use on lawns and parks. In one town in Quebec, an influential doctor saw the health effects of pesticides in her patients and rallied townspeople (meso) until the town banned (macro) their use. Today, 170 towns in Canada have similar laws (macro level).[56] About 80 percent of Canadians now live in places that restrict pesticide use. In the first year after Toronto's ban, pesticide use dropped by 88 percent.

COLLECTIVE PROBLEM, COLLECTIVE CITIZENS

According to Amitav Ghosh, we won't act on climate change if we leave it to individuals to make necessary changes because climate change is a collective predicament,[57] a classic collective action problem. Sufficient collective action is needed because only some individuals, businesses, and governments will voluntarily participate and reduce carbon emissions.

At a time when the collective is crucial to slow the climate crisis, Ghosh argues that in the dominant culture "the collective" has been exiled from politics, economics, and even literature. Highly individualistic cultures lack participation in civic life, whether voting or talking with neighbors. It's easy to think individually and act to enhance self-interest. Yet, it's entities with a high degree of social cohesion that will move us fastest toward climate resiliency. Social cohesion is important for responding to disaster and coping with it; strong social bonds, connections, and networks at the local level play an important role.

In pockets around the globe, collectives are emerging to address the climate crisis. In *Rising: Dispatches from the New American Shore*, Elizabeth Rush describes what sea-level rise and coastal flooding (affecting 500,000 Americans) can teach us about fighting climate change.[58] She noticed numerous flood-survivor groups popping up to help with immediate pre- and post-disaster needs. But when one flood became four or five, groups of citizens began looking at underlying causes and the need for bigger solutions.

Higher Ground is the largest coalition of flood survivors in the country (50,000 members); in addition to connecting frontline communities to pro-bono legal and scientific advice, the group has shifted focus from the singular to the systemic. They disrupted city council meetings by unfurling banners that read "The United Flooded States of America." Member groups filed lawsuits against unlawful wetland development and informed people about things that exacerbate flooding such as building on floodplains and climate change. Some member groups have advocated for "managed retreat" from flood-prone areas. Rush said her hope comes not from changed consumer habits but from people who recognize that their vulnerability to risk is *shared* and then come together to build coalitions seeking serious solutions. From this place, she says, it's easier to identify the root of the problem: "unjust development and the fossil fuel industry have made our communities less safe while super-charging the storms to come."[59]

One scholar said, "If the Anthropocene [human epoch] tells us anything, it is that the Age of the Individual is over."[60] Our climate change transition is simultaneously an individual journey and a group effort requiring massive cooperation and co-creation.

Bye Bye Bystander-ism

If you have *ever* felt fear for a climate-changed world, worried about the future of younger generations, or mourned what and who is being lost or harmed, it's time to get comfortable with your own power. Say goodbye to bystander-ism. Here are several ways to get started in your role as a social actor.

First, envision yourself a citizen in a collective, which includes all your existing social groups. Small consumer-oriented actions are at best a bridge to motivate you to participate in ways that make a more significant difference. Seek out important social others and share your worries, fear, ideas, and passion about the climate crisis. Consider what it means to belong to a collective —on your block, your city, your planet—that must cocreate our future.

Another important task is to grow your self-efficacy—believing that you can help address climate change, and feeling like you can bring it up in your social groups. It might be a different role for you, but you also might be surprised at the social support you receive.

A lack of self-efficacy can be a stumbling block for many who are not sure what they can do about climate change. Here are four suggested ways to bolster your self-efficacy.[61]

- First, by mastery experiences: performing a task successfully, like bringing up climate change in a social group. The second discussion is always easier, and so on, so continue toward mastery.
- Second, through social modeling: seeing someone like yourself succeed in climate change action shows you what that looks like. Then, you speak up and become a social model for others.
- Third, through social persuasion: being encouraged and believing you can succeed. This involves both outer support (which I hope you're getting here and from others) and inner resolve.
- Fourth, with a positive psychological response: achieved in part by working through the emotional states that convince you cannot succeed (the focus of the next chapter).

Your final task is to make climate change a *social reality* for you and those around you. As UK's Climate Outreach concluded, "[F]or most people, climate change is a scientific but not yet a 'social reality.' Most people have not yet heard a story about climate change that sounds like it was written for them, in language that connects with their interests, values, or identity."[62]

In your social group, present climate change as a story that's written for them and connects with norms and values. Research local impacts (warm winters, a struggling industry, heat and health). Connect local news headlines, such as pollution from climate-linked wildfires. Share personal stories about how climate change is affecting your group—right here, right now.

Then, act on that social reality. Here are examples of how three (imagined) social groups might take action *as a group* on the social reality of climate change: a family unit, an outdoor group, and a workplace. The actions target how the climate crisis affects what binds each group together. A key criterion for action: things that *greatly reduce carbon emissions* at a collective level, that address barriers and constraints for doing so, or that help the climate crisis become a true social reality. Think BIG. Even though some actions are undertaken by individuals, they involve collective participation and could serve as a bridge to larger actions.

The easy part is that group communication channels are already established. You have contact information, perhaps set gathering dates and times, and of course shared experiences and values. After a conversation or two on the social reality of the climate crisis, set aside some time to talk about ways the group can take action.

Family Group

Your "family" could consist of an extended family of origin, your immediate family, or what you consider your adopted family. After conversations that establish the social reality of the climate crisis for your family, brainstorm action to take as a family such as these suggestions.

- Share stories about individuals involved in climate action who inspire you like teen activist Greta Thunberg and others.
- Volunteer as a family for a citizen science project to help scientists better understand climate impacts. Count birds if you have a backyard feeder, note bud-burst for plants and trees,[63] or help a scientist count species or measure water and temperatures.
- Engage in citizen action as a family—march for climate action, march for clean air, march for the Eairth. Grassroots actions demonstrate citizen involvement, and it's fun—paint a sign, sing and chant, be outside, march with all those other people and dogs.
- Check with local museums or organizations for area field trips that observe changes from a shifting climate, whether a watershed, a farm, a forest, or a wetland.

- Attend a meeting of a local climate group (such as Citizens Climate Lobby), or a city or legislative meeting, especially when topics concerns climate change, air quality, EV charging stations, tax breaks for alternative energy, energy efficient buildings, and so on. Learn how citizens can participate in the democratic process.
- Use your family's interests to help determine where climate action is needed. If you're sports fans, see if your favorite team has addressed its carbon emissions; an array of athletes now speak out about the effects of climate change on sport and outdoor recreation.[64] If you're active in your church, use what your faith's doctrine says about care of the earth to broach the topic of climate change with fellow congregants.

Outdoor Enthusiasts

Let's say your group consists of ten close friends who for years have enjoyed outdoor activities of all kinds together: skiing, canoeing, hiking, and camping. Everyone is aware of climate change, but thus far your group hasn't taken the discussion beyond hand-wringing.

- At a minimum, organize carpools to outdoor excursions.
- Become active in a group that works to protect public lands for recreation or takes action on climate change. Snow-specific examples are POW (Protect Our Winters) and SOS (Save Our Snow).
- Invite those unfamiliar or unexperienced in the outdoors to join your adventures, Help someone who's disconnected from the living world get connected with it.
- Pressure lodges, outfitters and guides, and suppliers you depend upon to significantly reduce carbon emissions (beyond the "greenwashing" in which some engage[65]).
- Plan a "grounded experience" such a taking a field trip with a scientist who studies snow, or a government official who manages public lands.
- The outdoor gear industry promotes prodigious consumption, and virtually all its goods are made with petroleum products. Organize equipment swaps or pressure manufacturers to take-back gear to recycle or repurpose. At events, make sure to connect climate change to consequences for outdoor recreation and needed actions.

Workplace Group

This group could be coworkers who know each other well and perhaps socialize outside of work. Workplaces consume a lot of energy—electricity and heating and cooling, but also in the products they purchase, make, and sell.

Initiate a meeting with coworkers to bring up your concerns about the climate crisis and brainstorm actions your employer could take.

- Initiate workplace-wide conversations about climate change, including how it affects employees and their families, as well as your business. Industries from the military to agriculture have undertaken long-term planning on climate change impacts to operations.
- Implement energy-saving work practices to be required of all employees. Make energy-wise choices for supplies; buy recycled products. Set heating temperature to sixty-eight and cooling to seventy-eight. Encourage others in your industry (or building) to follow suit.
- Make the lunchroom and all sponsored events "no-waste." Purchase dishes and flatware and a dishwasher for the lunchroom. Hire caterers who serve food in glass or ceramic containers and provide dishes and utensils that are washed instead of discarded.
- Propose fair but generous tele-commuting policies for appropriate employees. Examine air and ground travel for meetings and switch to phone- or video-conferencing where appropriate. Provide employees with mass transit passes. Discuss company carpool vans.
- Propose affordable on-site daycare to eliminate driving for drop-offs and pick-ups. A small workplace could pool daycare with other businesses in the building or office park.

"BOWLING" FOR SOCIAL ACTORS

In his bestseller *Bowling Alone: The Collapse and Revival of American Community*, Robert Putnam argues that a big reduction of in-person social intercourse has undermined active civic engagement that a democracy requires from its citizens.[66] He notes (like Ghosh) the lack of "a collective": people don't vote, and membership and volunteering in civic groups (such as parent-teacher and fraternal associations) and religious organizations has plummeted, except within older generations. He points to bowling: although the number of bowlers has increased, the number participating in bowling leagues has decreased—and gone with it, the social interaction and civic discussions that occur. He suggests that a primary cause of the erosion of US social capital and participation is technology "individualizing" people's leisure time. Putnam argues that consequences of a decline in social engagement can be more suicide, depression, crime, and other social problems. On the other hand, when social involvement is high, kids perform better in school, neighborhoods are safer, government is better, and people are happier and healthier.

The collective crisis of climate change is in many respects emblematic of current civic life: individual awareness of climate change is high, but tuning-out is rampant. It's time to redirect individual energy and emotion away from the guilt and shame of consumer actions that will never solve the crisis. Being a lone bystander is not an option in a climate-changing world, and, it's too emotionally difficult. Your most effective and powerful role is as an engaged social actor—however you define that. It could be working with a social group to identify changes at the meso and macro level that will help reinvigorate civic life. It could be working with citizens and consumers to prod legislators and corporations to enact changes that will stabilize the climate. It could involve manual labor or writing or fundraising or talking, whatever your skills and abilities. The important part is not just to ask, "What can I do?" but to turn outward to the social others in your life and ask, "What can *we* do"?

NOTES

1. Ryan Collins and Rachel Adams-Heard, "Analysis|Flaring, or Why So Much Gas Is Going Up in Flames," *Washington Post*, September 4, 2019, https://www.washingtonpost.com/business/energy/flaring-or-why-so-much-gas-is-going-up-in-flames/2019/09/04/f3db2166-cf1b-11e9-a620-0a91656d7db6_story.html.

2. Matthew R. Johnson and Adam R. Coderre, "Opportunities for CO2 Equivalent Emissions Reductions via Flare and Vent Mitigation: A Case Study for Alberta, Canada," *International Journal of Greenhouse Gas Control* 8 (May 1, 2012): 121–31, doi:10.1016/j.ijggc.2012.02.004.

3. Seth Wynes and Kimberly A. Nicholas, "The Climate Mitigation Gap: Education and Government Recommendations Miss the Most Effective Individual Actions," *Environmental Research Letters* 12, no. 7 (July 2017): 074024, doi:10.1088/1748-9326/aa7541.

4. Individuals' home-based consumption constitutes a very small portion of resource use, as little as 25 percent of total energy usage, 10 percent of total water use, and 3 percent of municipal waste (Derrick Jensen, "Forget Shorter Showers: Why Personal Change Does Not Equate Political Change," *Orion*, August 2009). Of course, the resources used to produce food, laptops, and bath-towels don't get put in the individual's energy-use column, though our collective consumption drives these processes.

5. Stuart Capstick, Irene Lorenzoni, Adam Corner, and Lorraine Whitmarsh, "Prospects for Radical Emissions Reduction through Behavior and Lifestyle Change," *Carbon Management* 5, no. 4 (2014): 429–45.

6. Maxine Joselow, "Emissions: Quitting Burgers and Planes Won't Stop Warming, Experts Say," *E&E News*, December 6, 2019, https://www.eenews.net/stories/1061734031.

7. Bjorn Lomberg, "Your Electric Car and Vegetarian Diet Are Pointless Virtue Signaling in the Fight against Climate Change," *MarketWatch*, December 28, 2019,

https://www.marketwatch.com/story/your-electric-car-and-vegetarian-diet-are-pointless-virtue-signalling-in-fighting-climate-change-2019-12-26.

8. Michael E. Mann, "Lifestyle Changes Aren't Enough to Save the Planet. Here's What Could," *Time*, September 12, 2019, https://time.com/5669071/lifestyle-changes-climate-change/.

9. Lomberg, "Your Electric Car and Vegetarian Diet Are Pointless."

10. Emma Marris, "How to Stop Freaking out and Tackle Climate Change," *New York Times*, January 10, 2020.

11. A. Duit, "Patterns of Environmental Collective Action: Some Cross-National Findings," *Political Studies—Oxford* 59, no. 4 (2011): 900–20.

12. "China to Ban Bags and Other Single-Use Plastic," *BBC News*, January 20, 2020, sec. China, https://www.bbc.com/news/world-asia-china-51171491/.

13. P. Wesley Schultz, Jessica M. Nolan, Robert B. Cialdini, Noah J. Goldstein, and Vladas Griskevicius, "The Constructive, Destructive, and Reconstructive Power of Social Norms," *Psychological Science* 18, no. 5 (2007): 429–34.

14. Overall, an electric car produces far fewer emissions. However, if the electricity used to charge the car comes from coal or natural gas, the emissions still occurred – just someplace else.

15. Ezra Markowitz, Caroline Hodge, and Gabriel Harp, *Connecting on Climate: A Guide to Effective Climate Change Communication* (New York, NY: ecoAmerica and the Center for Research on Environmental Decisions, Columbia University, December 2014); Adam Corner and Jamie Clarke, *Talking Climate: From Research to Practice in Public Engagement* (Cham: Springer International Publishing, 2016); "Yale Program on Climate Change Communication: Messaging," *Yale University*, https://climatecommunication.yale.edu/topic/messaging/.

16. *Intergovernmental Panel on Climate Change*, "Climate Change Threatens Irreversible and Dangerous Impacts, But Options Exist to Limit Its Effects," 2014, http://www.un.org/climatechange/blog/2014/11/climate-change-threatens-irreversible-dangerous-impacts-options-exist-limit-effects/.

17. M. C. Nisbet and D. A. Scheufele, "What's Next for Science Communication? Promising Directions and Lingering Distractions," *American Journal of Botany* 96, no. 10 (2009): 1767–78; Patrick Sturgis and Nick Allum, "Science in Society: Re-Evaluating the Deficit Model of Public Attitudes," *Public Understanding of Science* 13, no. 1 (2004): 55–74.

18. Susanne C. Moser, "More Bad News: The Risk of Neglecting Emotional Responses to Climate Change Information," in *Creating a Climate for Change: Communicating Climate Change and Facilitating Social Change*, ed. Susanne C. Moser and Lisa Dilling (Cambridge: Cambridge University Press, 2007), 64–80; Lorraine Whitmarsh, Saffron O'Neill, and Irene Lorenzoni, eds., *Engaging the Public with Climate Change: Behaviour Change and Communication* (London: Earthscan, 2010).

19. LeeAnn Kahlor and Sonny Rosenthal, "If We Seek, Do We Learn?," *Science Communication* 30, no. 3 (2009): 380–414.

20. Paul M. Kellstedt, Sammy Zahran, and Arnold Vedlitz, "Personal Efficacy, the Information Environment, and Attitudes Toward Global Warming and Climate Change in the United States," *Risk Analysis* 28, no. 1 (2008): 113–26.

21. Jon Krosnick, Allyson Holbrook, Laura Lowe, and Penny Visser, "The Origins and Consequences of Democratic Citizen's Policy Agenda: A Study of Popular Concern about Global Warming," *Climatic Change* 77 (2006): 7–43.

22. Milena Buchs, AbuBakr S. Bahaj, Luke Blunden, et al., "Promoting Low Carbon Behaviours through Personalised Information? Longterm Evaluation of a Carbon Calculator Interview," *Energy Policy* 120 (2018): 284–93.

23. Matthew J. Hornsey, Emily A. Harris, Paul G. Bain, and Kelly S. Fielding, "Meta-Analyses of the Determinants and Outcomes of Belief in Climate Change," *Nature Climate Change* 6, no. 6 (2016): 622–26.

24. Aaron M. McCright and Riley E. Dunlap, "The Politicization of Climate Change and Polarization in the American Public's Views of Global Warming, 2001–2010," *Sociological Quarterly* 52, no. 2 (2011): 155–94; A. McCright, R. Dunlap, and C. Xiao, "Increasing Influence of Party Identification on Perceived Scientific Agreement and Support for Government Action on Climate Change in the United States, 2006–12," *Weather Climate and Society* 6, no. 2 (2014): 194–201.

25. Riley E. Dunlap, "Lay Perceptions of Global Risk Public Views of Global Warming in Cross-National Context," *International Sociology* 13, no. 4 (1998): 473–98.

26. John Cook, John Kotcher, Neil Stenhouse, and Ed Maibach, "Editorial: Public Will, Activism and Climate Change Communication," *Frontiers in Communication* 4, no. 72 (November 29, 2019), doi:10.3389/fcomm.2019.00072.

27. Eviatar Zerubavel, *The Elephant in the Room: Silence and Denial in Everyday Life* (New York, NY and Oxford: Oxford University Press, 2008), 47.

28. Kari Marie Norgaard, *Living in Denial: Climate Change, Emotions, and Everyday Life* (Cambridge, MA: MIT Press, 2011), 5.

29. Andrew Hoffman, *How Culture Shapes the Climate Change Debate* (Stanford, CA: Stanford University Press, 2015), 16.

30. Malcom Gladwell, "Why the Revolution Will Not Be Tweeted," *The New Yorker*, October 4, 2010.

31. Maria Knight Lapinski and Rajiv N. Rimal, "An Explication of Social Norms," *Communication Theory* 15, no. 2 (2005): 127–47.

32. Ann P. Kinzig, Paul R. Ehrlich, Lee J. Alston, et al., "Social Norms and Global Environmental Challenges: The Complex Interaction of Behaviors, Values, and Policy," *BioScience* 63, no. 3 (2013): 164–75.

33. Robert B. Cialdini, "Descriptive Social Norms as Underappreciated Sources of Social Control," *Psychometrika* 72, no. 2 (2007): 263–68.

34. Robert B. Cialdini, "Crafting Normative Messages to Protect the Environment," *Current Directions in Psychological Science* 12, no. 4 (2003): 105–9.

35. A. Leiserowitz, E. Maibach, S. Rosenthal, J. Kotcher, P. Bergquist, M. Ballew, M. Goldberg, and A. Gustafson, *Climate Change in the American Mind: November 2019* (New Haven, CT: Yale Program on Climate Communication and George Mason University, 2019).

36. Gregory Owen Thomas, Wouter Poortinga, and Elena Sautkina, "The Welsh Single-Use Carrier Bag Charge and Behavioural Spillover," *Journal of Environmental Psychology* 47 (2016): 126–35.

37. A. Miyashita, "Where Do Norms Come from? Foundations of Japan's Postwar Pacifism," *International Relations of the Asia-Pacific* 7, no. 1 (2007): 99–120, 99.

38. Candis Callison, *How Climate Change Comes to Matter: The Communal Life of Facts* (Durham, NC: Duke University Press, 2015).

39. Kate Kenski, Kathleen Hall Jamieson, and Natalie Jomini Stroud, "Selective Exposure Theories," in *The Oxford Handbook of Political Communication* (Oxford: Oxford University Press, 2017).

40. Douglas Blanks Hindman, "Mass Media Flow and Differential Distribution of Politically Disputed Beliefs: The Belief Gap Hypothesis," *Journalism & Mass Communication Quarterly* 86, no. 4 (2009): 790–808, p. 790.

41. McCright and Dunlap, "The Politicization of Climate Change," 155.

42. Norgaard, *Living in Denial*, 6.

43. Arlie Russell Hochschild, *The Managed Heart: Commercialization of Human Feeling* (Berkeley, CA: University of California Press, 2012), 223.

44. Norgaard, *Living in Denial*, 5.

45. Norgaard, *Living in Denial*, 11.

46. Stanley Cohen, *States of Denial: Knowing about Atrocities and Suffering* (Cambridge, UK: Polity, 2015), 9.

47. Z. Leviston I. Walker, and S. Morwinski, "Your Opinion on Climate Change Might Not Be as Common as You Think," *Nature Climate Change* 3, no. 4 (April 2013): 334–37.

48. Jenny Onyx and Paul Bullen, "Measuring Social Capital in Five Communities," *Journal of Applied Behavioral Science* 36, no. 1 (March 2000): 23–42.

49. Corner and Clarke, *Talking Climate*, 80.

50. Norgaard, *Living in Denial*, 12.

51. Norgaard, *Living in Denial*, 210.

52. Norgaard, *Living in Denial*.

53. Norgaard, *Living in Denial*, 210.

54. *U.S. Centers for Disease Control and Prevention*, https://www.cdc.gov/tobacco/data_statistics/fact_sheets/adult_data/cig_smoking/index.htm.

55. James Ellsmoor, "Climate Emergency Declarations: How Cities Are Leading The Charge," *Forbes*, July 20, 2019, https://www.forbes.com/sites/jamesellsmoor/2019/07/20/climate-emergency-declarations-how-cities-are-leading-the-charge/.

56. Keith Tyrell, "Pesticide-Free Towns and Cities: Citizen Power in Action," *Ecologist*, October 6, 2015, https://theecologist.org/2015/oct/06/pesticide-free-towns-and-cities-citizen-power-action.

57. Amitav Ghosh, *The Great Derangement: Climate Change and the Unthinkable* (Chicago, IL: University of Chicago Press, 2016).

58. Elizabeth Rush, *Rising: Dispatches from the New American Shore* (Minneapolis, MN: Milkweed Editions, 2018).

59. Elizabeth Rush, "What 500,000 Americans Hit by Floods Can Teach Us about Fighting Climate Change," *The Guardian*, August 30, 2019, https://www.theguardian.com/commentisfree/2019/aug/30/what-500000-americans-hit-by-floods-can-teach-us-about-fighting-climate-change.

60. R. D. K. Herman, "Traditional Knowledge in a Time of Crisis: Climate Change, Culture and Communication," *Sustainability Science* 11, no. 1 (January 1, 2016): 163–76, p. 174, doi:10.1007/s11625-015-0305-9.

61. Albert Bandura, "An Agentic Perspective on Positive Psychology," in *Positive Psychology: Exploring the Best in People*, ed. S. J. Lopez (Westport, CT: Greenwood Publishing Company, 2008), 1:167–96.

62. Corner and Clarke, *Talking Climate*.

63. https://feederwatch.org/ and https://www.usanpn.org/nn/become-observer.

64. See *Sport4Climate Initiatives*, https://www.connect4climate.org/initiatives/sport4climate.

65. Julia B. Corbett, *Communicating Nature; How We Create and Understand Environmental Messages* (Washington, DC: Island Press, 2006), 274.

66. Robert D. Putnam, *Bowling Alone: The Collapse and Revival of American Community* (New York, NY: Simon & Schuster, 2000).

Chapter 4

Emotions and Climate Silence

In September 2018, I was evacuated for several weeks from a large, wind-whipped wildfire; the terror, dread, uncertainty, fear, and sadness I felt were intense and palpable. Two months later, I relived my trauma super-sized when fire erupted in Paradise, California. In 2020, when I read that rescuers heard koalas screaming from the trees during the Australian bushfires, the grief, loss, and profound sorrow flooded back and overwhelmed me. For months, I felt unmoored as I rode a wave between despair and hope.

Even if you haven't experienced climate change this acutely, you nevertheless do experience it emotionally. All of us—youth, climate activists, parents, climate scientists, and even those who claim they never think about it—struggle with the emotions of climate change and on some days want to turn away. You know climate change is a "super-wicked" problem, you're embedded in a fossil fuel culture that is not taking strong action, and you blame and shame yourself for not doing more. Every day, more scientific evidence and more climate consequences arrive at your doorstep and inbox. No wonder you experience strong, sometimes overwhelming emotions: anxiety, anger, sadness, despair, grief, guilt, loss, fear.

An acute inability to deal emotionally with climate change is contributing to inaction. Discomfort with (or ignorance of) these emotions exacerbates climate silence in social settings and the larger culture. Increasingly, psychologists realize that tackling burgeoning "climate anxiety" and tackling the climate crisis are intrinsically linked. According to Oxford clinical psychologist Patrick Kennedy-Williams, "the cure to climate anxiety is the same as the cure for climate change: action. It is about getting out and doing something that helps."[1] And what helps on the eco-anxiety front is facing, working through, and talking about the difficult emotions you feel. Until you do, you are unlikely to be ready or receptive to hear and process much information or feel motivated to act.

Facing climate change and accepting it emotionally is not the same as acquiescence or resignation, of throwing up your hands and continuing life as usual. It's an acceptance that we are where we are, that it's scary, and that we have to contend with it for the rest of our lives.[2]

The new direction presented by this chapter is strategies for recognizing and communicating these emotions—to open the emotional door of climate change and go inside. Psychologists and medical doctors recognize the significant mental health tolls from climate change and the need to build emotional resilience. This is nurtured by understanding how people feel and make sense of the crisis, redirecting fears, and channeling energies. The chapter explores the various defense mechanisms we employ to avoid emotions, as well as mourning and grieving, building emotional resilience, and accepting a changed "normal."

A key standpoint that I adopt in this book—put forth by psychologist Renee Lertzman in *Environmental Melancholia*—is that people do indeed care and have concern for nature and the planet's ecosystems.[3] While many communicators strive to persuade an "apathetic" public to care and act on climate change, Lertzman maintains that people care a great deal but get caught up in complicated dilemmas and "psychic negotiations" that make it very hard to take action. Even those who care deeply about the Eairth may feel paralyzed and not know where to start.

THE MENTAL HEALTH TOLLS OF CLIMATE CHANGE

Mental health challenges are perhaps the most overlooked consequence of climate change.[4] The American Psychological Association (APA) reported that the mental health tolls of climate change are far reaching; stress, depression, and anxiety strain social and community relationships and are linked to increases in aggression, violence, and crime. Negative psychological responses to the climate crisis are increasing: conflict avoidance, fatalism, fear, helplessness, and resignation. The APA conclusion: "These responses are keeping us, and our nation, from properly addressing the core causes of and solutions for our changing climate, and from building and supporting psychological resiliency."[5]

You may experience mental health impacts as "chronic" (persistent anxiety) and/or "acute" (an intense, frightening experience). Climate-linked events such as floods, heat waves, wildfires, and hurricanes can cause immediate, acute psychological trauma. Survivors might have lost property or livelihood, suffered injuries, or lost people or animals they love. Those affected by Hurricane Katrina in New Orleans had increased levels of PTSD (post-traumatic stress disorder), suicide or thoughts of it, and depression.[6] Traumatic stress, terror, and shock occurred for those who had experienced flooding,[7] and trauma symptoms and PTSD remained for years for survivors of extreme bushfires.[8] Evacuees and refugees have higher rates of immediate and lifetime PTSD.[9]

A record 7 million people worldwide were displaced in the first half of 2019 by extreme weather[10] (which through event attribution can often be linked to climate change). In 2018 in the US, over a million people fled floods, two hurricanes, numerous wildfires, and other extreme weather events; I can attest to the stress and trauma my wildfire evacuation created. Being exiled from one's home place threatens your very sense of self, safety, and belonging. Worldwide, an ever-growing cadre of people possess personal experience with the mental health impacts of extreme climate-linked events.

Beyond severe events, for all of us, "global environmental change and regional ecological decline are increasingly embedded within everyday experience, evoking strong mental and emotional responses."[11] The "everyday experience" might be hotter temperatures (and smoggier skies), mild winters, or altered seasons. There is evidence (historical and current) that hotter temperatures are associated with an increased use of emergency mental health services and increased suicide;[12] when temperatures go up, so do aggression and violence.[13] Heat, record precipitation, and changes in your environment (such as air pollution) can reduce your sense of control over your life, which can lead to mental health impacts.[14]

As the APA states, "loss of place is not a trivial experience."[15] Your strong, emotional attachment to a place—a hometown, a river, a prairie—provides stability, security, and personal identity, and creates happiness and satisfaction. For those with place-dependent ways-of-life or occupations (farming, fishing, raising animals), your identity is inextricably bound to the land. If your farm fields flood, your grazing land or orchards burn, or your coastal waters warm and acidify, your occupational identity is greatly affected. Extreme depression and ecological grief have afflicted Canadian Inuits and Australian Wheatbelt farmers.[16] Some medical schools (such as Johns Hopkins, Yale, and University of California San Francisco) now train future doctors and health professionals to recognize and treat mental illness linked to the climate crisis. The Climate Psychiatry Alliance helps their profession and the public learn about the profound impacts on mental health and well-being caused by climate disruption.

Climate change also means serious mental health issues for climate scientists on the front lines of disturbing ecosystem change and who possess foreknowledge that others seemingly ignore. Some scientists experience deep grief and depression for what's being lost (coral reefs, forests, butterflies, or bats), anger against the inaction, and attacks from climate deniers.[17] One study examined the "emotional management strategies" used by climate scientists to cope with their daily immersion in the subject,[18] concluding that climate scientists might be the canaries in the psychological coal mine regarding the mental health tolls that await us all.

Stress, which we all experience at some time, is a physiological response that occurs when you don't believe you have the capacity to respond and adapt to a given situation. Not knowing how to respond or adapt to climate change is a widespread sentiment. Stress can include worry about the future and feeling vulnerable and helpless in the face of threats, and can be full of grief and despair. Chronic stress lowers immune system response, affects sleep, and increases the stress hormone cortisol, which if prolonged affects digestion and memory. The mental health tolls of the climate crisis are significant, and millions have already experienced them deeply. To avoid the increasing fatalism, helplessness, and resignation noted by the APA, we need ways to face our emotions and strengthen emotional resiliency.

FACING EMOTIONS AND FEARS

Emotions are present even when a statement or situation doesn't appear to have much to do with emotions. Consider, for example, these viewpoints.

"Sure, the climate is changing, but humans aren't responsible for it—it's natural."

"I know climate change is real, but day-to-day, I just can't deal with or think about it."

"Climate change is a hoax."

"Climate change is really scary. I worry constantly about it, but I feel pretty paralyzed."

Even though only one statement uses emotion words (*scary*, *worry*), what do these four viewpoints—which seem worlds apart—hold in common? Each one is motivated in large part by the person's feelings, worries, and defenses—unconscious and conscious—about what the climate crisis means for his or her life.

For all individuals—regardless of viewpoint—the emotions of climate change are hard to navigate. The crisis is enormous and often unfathomable; it taps base fears and anxieties (often at an unconscious level). This triggers defense mechanisms to rationalize or hold the issue at bay and makes us unreceptive to thinking, talking, or engaging as social actors. According to psychologists, "denial" (expressed in two statements above) is both a powerful defense mechanism in response to conflicted emotions and the first stage in the grieving process.

Existing discourses on the climate crisis tend to be highly rationalized with very little space for emotions, even though they are fully present. The most dominant emotions associated with climate change are fear, feeling frustrated and overwhelmed, anger, guilt, and shame.[19] Emotions, negative and positive, serve as interpreters and translators of the events in our lives and help drive our behaviors. You know from many problems—such as a relationship conflict or a new medical condition—that until you acknowledged or faced your emotions and fears, it was difficult to reconcile or move forward. The same is true for emotions about the climate crisis.

When creatures encounter a threat, their immediate behavioral choices are "fight, flight, or freeze," meaning to remain motionless. But even if you "freeze" (an ironic choice in a warming world), you internalize the threat: heart rate and breathing increase, and your mind races. You cannot control this; your body is hardwired to respond when you are afraid. The response may not be heart-pounding but instead chronic, ambient anxiety.

In a Survey of American Fears, climate change was listed in the top ten things Americans were "afraid" or "very afraid" of.[20] Five of the top ten concerns were environmental threats: air pollution, water pollution, pollution of bodies of water, climate change, and extinction. The researchers said policy changes were likely factors in the 2017 and 2018 surveys (President Trump weakened over a hundred environmental protections and withdrew from the Paris Climate Accord). It's interesting that although few Americans list climate change as one of the most important issues when asked by opinion pollsters,[21] over half surveyed here said they *feared* it.

Young people are especially vulnerable to eco-anxiety and fear. A recent survey of US youth (commissioned by the *Washington Post* and Kaiser Family Foundation) found that about 57 percent reported that climate change makes them feel afraid.[22] More than 70 percent believe climate change will cause a moderate or great deal of harm to people in their generation.

Communicators (both academic and professional) don't agree whether "fear appeals" in messages are effective or counterproductive.[23] Perhaps that is beside the point if so many adults (and youth) already possess considerable fear about climate change. Psychologists and counselors argue that instead of ignoring or avoiding emotions like fear that surround the climate crisis, emotions should be addressed directly and fully. Neglecting the emotional reception to climate change messages makes communication and outreach more likely to fail.[24]

At the same time, positive emotions (like love or hope) can be important catalysts for engagement—such as love of a place or a creature. Climate scientist and communicator Katharine Hayhoe recommends engaging others not just with your head but with your heart, voicing why you care about climate change.[25]

There is a tendency to perceive science as "rational" and divorced from emotion. Yet neuroscience has discovered that emotions are not a threat to rational deliberation about climate change. Quite the opposite: emotions are a necessary source of reflection and insight concerning the moral impact of climate change and our ability to make good decisions.[26] Neuroscientist Antonio Damasio has shown that purely rational beings cannot make proper practical assessments without emotions, especially when it comes to moral judgments.[27] Purely rational beliefs can be misleading, and emotions can correct them. (Remember that Dr. Spock on *Star Trek* often sought emotional input from others to complement his solely rational mind.) Emotional engagement has been found to lead to a higher degree of motivation than a detached, "rational" stance on climate change, particularly when the message moves beyond "numbing numbers"[28] to feelings of justice, sympathy, and compassion about people and events close by.

Even if the typical images of climate change (blazing wildfires, heat-rippled landscapes, or flooded cities) don't seem personally relevant or dangerous to you, the images nevertheless arouse anxiety (sometimes profoundly). That can create *cognitive dissonance*—the mental discomfort you feel when you receive information that contradicts or is contrary to beliefs or ideas you hold, such as thinking that climate change is a small and distant threat. Processing these images can also trigger a variety of defense mechanisms.

DEFENSE MECHANISMS WE EMPLOY

If you read about animals killed by wildfires, your conscious and/or unconscious mind may seek to avoid emotions that were aroused. In psychological terms, a defense mechanism protects a person from unwanted or painful ideas or emotions. A simple defense is simply forgetting what you read, or in some way rationalizing or intellectualizing the information. Often, a defense mechanism is deployed without your being consciously aware; it's a way your brain safeguards you from feelings or thoughts that are difficult for your conscious mind to cope with. However, if defense mechanisms distort reality and unconsciously allow you to avoid important, relevant emotions, that's not helpful for facing your emotions, including about climate change.

Psychological research tells us that for climate change, the limbic system (the complex system of nerves and networks in the brain that controls basic emotions and drives) strongly guides us. When emotions are driving the bus, the prefrontal lobe (implicated in abstract thinking and decision making) is essentially shut down by fear and anxiety.[29] This makes it easy to fall back into one's polarized position (and the information sources and social groups that reinforce that position) and employ emotion-based defense mechanisms. When anxieties are triggered, defenses are almost always inadvertently triggered.

Let's discuss six types of defense mechanisms: apathy, denial, disavowal, projection (or "othering"), ambivalence, and splitting.

The following statement communicates *apathy*, which many assume is a prevalent stance about climate change: people just don't care.

> *Sure, I've heard about climate change, but hey, it just isn't high on my list of things to worry about. I'm just living my life, ya know?*

But if you accept the opposite—that everyone does care for nature and ecosystems—what else might be going on here? Saying you don't care about climate change pushes aside painful anxieties and helps avoid conflicted feelings and thoughts about its threat. We all want to believe that our surrounding environment is healthy and that we're safe from the effects of climate change; information to the contrary makes us anxious and fearful. If you unconsciously adopt an apathetic stance, it seems to manage your fears and distance the issue. Apathy can be driven by a sense of helplessness, which for climate change is part actual (the threat is too large to face on your own) and in part manufactured (by projecting blame and responsibility onto others, such as industry or previous generations).[30]

One of the most common defense mechanisms is *denial*, seen in this statement:

> *The hand-wringing about climate change is so overblown. This is all just a normal fluctuation in the climate, which happens all the time.*

Climate denial is an extremely powerful emotional defense—it's a refusal to accept reality and recognize that something is occurring (like addicts who deny they have a problem). It's a bit like an ostrich effect; individuals are so driven to avoid learning about the science of climate change that they actively construct a safer alternative narrative.[31] This type of denial is very deeply held, and it takes an enormous investment of energy to maintain one's denial in light of the mountains of evidence and frequent, visible changes and disasters. (We'll return to the process of denial later in the chapter, where researchers have explored and tested its many forms.)

Only a step away from denial is *disavowal*, which is when a person does not support, acknowledge, or show responsibility for something that is happening:

> *I know climate change is happening but I'm going to keep doing what I'm doing anyway.*

When a lack of responsibility actively points fingers of blame at others, it's the defense mechanism of *projection*, sometimes called "othering":

If liberal environmentalists hadn't opposed logging, the big fires in California wouldn't have happened.

I've heard a half-dozen versions of this retort (applied to various fires in several states), which displaces blame (and anger) onto others. This projection denies that climate change played any role in the fires – although scientists concluded that climate change plays a far greater role in severe wildfires than either land management practices or past fire suppression.[32] Instead, blame is projected onto a targeted "other" or "enemy." By pointing a finger at environmentalists as responsible, it abrogates both personal responsibility and Pacific Gas & Electric's responsibility to clear flammables around its powerlines and maintain its equipment.

The defense mechanism of *ambivalence* refers to holding two mixed or contradictory emotions at the same time. Negotiating and integrating what we're feeling about the warming planet with our experiences and societal interrelationships is indeed a complex and seemingly contradictory process. This statement expresses emotions and paradoxes we all face:

It's so sad what's happening now with climate change—I mean all the fires, the weird weather, floods, it's horrible. It makes me afraid of what my kids will face. But what am I supposed to do? I can't exactly quit driving my car, and we don't want to mess up the economy.

This person is not denying climate change, so at first glance, this might not seem like a defense mechanism. But when you are ambivalent, your mixed or contradictory feelings point to a reluctance or inability to acknowledge the *source* of distress.[33] Let's examine what the statement reveals about our complex psychic negotiations and how those affect our ability to face and engage with climate change.

Our love-hate relationship with fossil fuels fits this well: we understand the harms and consequences of industrial-level reliance on them but can be reluctant to separate and feel fused to the objects they create (whether big-screen TVs or gas grills). Ambivalence toward the environment is related to our sense of security, both physical and emotional (such as by pointing to "jobs" or "needed" goods). At the same time, the duality of love-hate feelings in the face of environmental loss can lead to unconscious guilt, which a person both internalizes and externally attaches to desired objects (designer jeans or expresso makers).

The presence of ambivalence, Lertzman says, is bound up with feelings of guilt and conflict regarding industry and the environment. If one doesn't name and come to terms with this distress, the sense of loss and disappointment remains directed inward and unresolved. A more constructive and appropriate response (and use of emotional energy) would be to mourn the loss and even express anger in a way that supports action or reparation: *"I am so angry that this country is not taking the climate crisis seriously, and I'm so mad that my city isn't aggressively expanding mass transit!"*

The final defense mechanism *splitting* divides objects into good and bad qualities:

This is a great place to live, but the wildfire smoke is bad every summer now.

Internally, splitting disassociates us from parts of ourselves that are too threatening or overwhelming—perhaps here a fear of wildfire or how the hometown has been changed by climate change. Splitting compartmentalizes the good versus bad—such as thinking that current industrial practices are a "necessary evil" and, therefore, acceptable in the ways they harm us and the living world (which contributes to feeling stuck in fossil fuel culture).

Compartmentalizing dissociates us from our interrelatedness with Eairth systems. Our culture tends not to recognize environmental loss as valid,[34] even though threats to our environment dissolve things we can feel certain about (normal seasons or personal health) and rupture our sense of who we are and what it means to be human.[35] Indeed, Eairth is not only the foundation of our very lives, the living world is also one of the most basically important ingredients of human psychological existence;[36] we are connected at a very deep level whether we are conscious of that or not.

WORKING THROUGH EMOTIONS

Even if the statements presented above seem innocuous, Lertzman and other psychoanalysts conclude that the emotions underlying such defenses impede our capacity to face climate change and take action. I totally understand not wanting to "go there," of being reluctant to face pain and feel loss. At times, it can seem easier to invest lots of energy to keep feelings at arm's length instead of attending to them. But as you probably know from past experience, there is much to learn from uncomfortable feelings if they are recognized and expressed.

Negotiating and processing emotions also can help people engage more authentically and personally in an issue and explore ways to contribute to solutions. From the mid-1980s to mid-1990s, the AIDS quilts became public memorials to remember those who died and a catalyst for mobilization. In each quilt square, an individual lost to AIDS was eulogized; squares were sewn into enormous quilts and displayed in public places. It's just as true for the climate crisis that facing loss is both a necessity of life and a call to responsibility, accepting past losses and losses to come.[37] Our collective pain over the losses and injustices of climate change can unite and motivate us, encouraging ethical and political responses and actions.

Communicators can help create ways to speak about, name, and process our emotions about climate change and integrate them into daily life. A simple and free method is to *talk* with close friends or family with whom you feel safe and able to open up. Even though emotions feel individual, they are very much shared, circulated, and even contagious.[38]

If you doubt the power of simply talking about climate change, consider this. You already use conversation as a key way to process emotions and events. You have likely experienced a time when until you began talking about something difficult, you were not fully aware of the emotions underlying it – that is, you were unconscious and not self-aware of how you felt. Thus, opportunities and safe spaces to converse and the permission to express what we feel are key; we need acknowledgment that our feelings are okay and normal. When you meet someone in a conversation where he is *now* with his emotions, it disarms and softens the tendency to defend and distance.[39] Box 4.1 has questions to help explore your emotions about climate change.

Lertzman encourages *empathy* when engaging with others.[40] When people fear for their security and an unknown climate-changed future, what's the best response? Not a prescriptive message or lesson about climate change, but empathy and asking people what they need, right now, to feel safer and more secure. If you recall a recent disaster, this was likely the process that unfolded. The community "came together," people stepped in to help others with immediate needs and safety, and feelings were fully out in the open and expressed.

A final way to deal with emotions about climate change: become involved in a collective action. Psychologists and teachers noted that participation in the UK school climate strikes reduced symptoms of mental ill-health among young people.[41] Activism helped counteract their powerlessness, anger, and disenchantment and heightened a sense of agency as part of the global youth-led movement. The youth returned to school "buoyed up" and determined.

These are important lessons for communicators; acknowledge existing emotions and fears (and provide space and opportunities to express them) rather than avoiding them.

Box 4.1   Working through Emotions about Climate Change

It's no surprise that examining your emotions is hard! True learning always requires some internal conflict or dissonance. Looking deep inside requires you to tolerate discomfort and face your resistance, defenses, and compulsions. It helps to be curious and willing to investigate how you navigate your emotional self. As you progress through the questions, allow time to reflect. You can contemplate the questions below by yourself, but I recommend talking through them with others with whom you feel safe and able to open up. Mourning is both personal and social, where we acknowledge, recognize, and share loss and sadness.

A guiding mantra of sorts for a group discussion of climate change emotions can be, "We know this is scary and overwhelming, but we are in this together. We are where we are, and we have to contend with climate change for the rest of our lives." Strive for honest, mature conversations that allow people to express what they feel but are hesitant to voice, such as disappointment, anger, outrage, despair, and sadness. Remember to listen as much (if not more) as you speak and remain comfortable with silences.

- What care, concern, and love do I have for the living world?
- How does climate change make me feel? Are any emotions missing or less conscious? Why is that?
- What emotions and thoughts do I have about the climate crisis that make me most afraid or cause me the most distress?
- Do I talk about climate change with important others in my life? Why or why not? Are there fears or beliefs that hold me back?
- What ecological losses have I experienced personally? (This is different from distant losses you see on the news.) If I'm not aware of experiencing any personal loss, why is that?
- What do I need to mourn? (Mourning is learning to recover from a loss, a transition that is mental, emotional, and personal.)
- Where am I in Kübler-Ross' five stages of grief? (Denial, Anger, Bargaining, Depression, Acceptance) Is there one stage where I get stuck?
- Is there a way I come across as apathetic about climate change? If so, are there underlying emotions or thoughts that trigger that?
- What ambivalence do I hold about my lifestyle regarding the climate crisis? Can I identify emotions or fears behind my ambivalence?
- How do I split and compartmentalize climate change? Do I have any trouble reconciling its good/bad qualities? If so, why is that?
- How do I project the problem of climate change onto others or external things?
- What kinds of images, information, and sentiments prompt my defense of the status quo? What do I fear or resist about changing the status quo to address climate change?
- Which types of needs (epistemic, existential, or relational) get triggered for me with news of climate change? (Defined below in System Justification Theory.)
- What exactly do I fear about the future? Do I fear or resist certain sacrifices or losses? How might I reimagine those in positive, solution-oriented ways?

ECOLOGICAL GRIEF AND MOURNING

Take another look at the earlier statement of ambivalence and the emotions expressed.

> *It's so sad what's happening now with climate change—I mean all the fires, the weird weather, floods, it's horrible. It makes me afraid of what my kids will face. But what am I supposed to do? I can't exactly quit driving my car, and we don't want to mess up the economy.*

Lertzman calls this *environmental melancholia*, when those who care deeply about the well-being of the living world feel paralyzed and unable to translate their concern into action, which includes unresolved mourning of a loss or anticipated loss.[42] This person's anxiety is palpable: she is aware of her sadness and the destruction, and she anticipates losses her kids will face. A sense of loss takes many forms: loss of identity, of lifestyle, and of environmental "normal."

It's hard to feel grief and loss when those feelings are partitioned away or projected elsewhere. Psychotherapist (and Director of Cambridge Carbon Footprint) Rosemary Randall says that in traditional "problem" narratives about climate change, loss is a dominant theme—loss of healthy oceans, species and habitats, crops and water, and illness, often portrayed in dramatic images.[43] But loss in these narratives can nevertheless feel unreal, remote, and in the future. Meanwhile, "solutions" narratives mention nothing about loss, and instead focus on small steps, "green consumerism," and how technology will save us. Randall says problem narratives and solutions narratives run parallel to each other but remain *unconnected*. Loss and immediacy are split away, blame is projected onto others, and people are protected from the need to truly face and mourn the losses associated with climate change.

Think of messages where unconnected but parallel narratives occur. Problem narratives of climate-related wildfires give statistics of acres burned, life lost, and homes burned, but rarely mention climate change or steps to take. Meanwhile, solutions narratives about "carbon capture" technology don't mention the consequences and losses that excess carbon creates. Unconnected parallel narratives are common in news stories, but also occur in casual conversations and political speeches: personal feelings of loss are rarely mentioned, aseptically scrubbed away. A more integrated communication response would acknowledge both the losses (and emotional costs) and positive future steps we can take.

Most often, we associate grief and mourning with losing a fellow human. Rarely are these concepts extended to losses we encounter in the nonhuman world,[44] making ecological grief a "disenfranchised grief" not publicly or openly acknowledged, even though intense feelings are a natural, legitimate response to ecological loss.[45] Losing a forest or beloved fish species really hurts, and it feels like I lost a part of myself. "Ecological grief" makes sense when we recognize our interdependency with ecological systems and our ethical responsibilities toward them.

Grief is the internal response to loss that happens in your body and heart; *mourning* is the period when you learn to live again after a loss, a transition that is mental, emotional, and personal, and sometimes endures for a very long time. In mourning, "We not only lose something that was loved, but we also lose our former selves, the way we used to be before the loss. We are changed internally and externally by the loss in ways that we cannot predict or control."[46] Ecological grief is associated with three kinds of loss: physical ecological loss (the disappearance or degradation of species or entire ecosystems), loss of environmental knowledge systems and identity (lost ability or confidence to live upon or work in the living world), and grief around anticipated future losses.[47]

Philosopher Judith Butler recognizes that grief and mourning have "we-creating" capacities; they help us recognize our common bonds and shared vulnerability, which can bring us together and lead to political and ethical transformation.[48] Ashlee Cunsolo, a social scientist and health researcher in northern Canada, maintains that "public mourning can be an important mechanism for political mobilization, the counteraction of dominant discourses . . . and for sharing the grief experienced from climatic and environmental change."[49] Extinction Rebellion activists effectively used public mourning in civil disobedience protests when they dressed in mourning clothes and carried small coffins to represent species and ecosystem loss.

If you have experienced a profound loss, you know there is no one clear path to move you through your grief; grief is a nonlinear process that individuals navigate in unique ways. Nevertheless, scholars and practitioners have created models that help you understand the process of grieving, including for climate change.

The most recognized model is seminal grief theorist Elisabeth Kübler-Ross' five stages of grief.[50] Note that some steps are tied to defense mechanisms (the example statements are mine.)

- Denial (*Climate change is a big hoax.*)
- Anger (*The enemy is Big Oil, politics, and industrial capitalism.*)
- Bargaining (*I'll recycle and buy a hybrid car—is that enough?*)
- Depression (*What's the point; there's nothing we can do.*)
- Acceptance (*Climate change is here and it's real, and I know it has great consequences for my life and all life, now and in the future.*)

These stages are not necessarily linear and a person may cycle back and forth through them. And, "acceptance" isn't feeling that climate change is "okay"

> **Box 4.2 Good Grief's Ten Steps to Personal Resilience in a Chaotic Climate**
>
> 1. Accept the Severity of the Predicament
> 2. Acknowledge That I Am Part of the Problem and Solutions
> 3. Practice Being with Uncertainty
> 4. Honor My Own Mortality and the Mortality of All
> 5. Do Inner Work
> 6. Develop Awareness of Brain Patterns and Perceptions
> 7. Practice Gratitude, Appreciate Beauty, and Create Meaning
> 8. Take Breaks and Rest
> 9. Show Up
> 10. Reinvest into Meaningful Efforts
>
> (Source: https://www.goodgriefnetwork.org)

but recognizing that the new reality is the permanent reality—both in your personal and professional life.[51]

Psychologist and grief counselor William Worden's four tasks of grief[52] are adaptable to climate change. Worden suggests that the work of grief is a series of tasks that can be embraced or refused, and tackled or abandoned. The first task is accepting the reality of losses from climate change (which can take years and requires moving beyond intellectual acceptance). This is followed by tasks of working through painful emotions, adjusting to the new environment, and reinvesting and redirecting emotional energy (previously diverted by depression and grieving).

A development model of grief by psychologists Robert Neimeyer and Joanne Cacciatore portrays grief as moving through three phases (not necessarily in linear fashion): reacting, reconstructing, and reorienting.[53] I found this model useful after my wildfire evacuation and return. When I first saw the monotone of black, charred land and skeleton trees around my house, I experienced intense and profound grief. It's been a long process of trying to reconstruct my life there and reorient to a landscape that is slowly recovering but is forever changed.

Finally, Good Grief in Salt Lake City (with chapters in additional locations) holds monthly gatherings to work through the "heavy feelings" of climate change.[54] Founder Laura Schmidt said her motivating question was "How do we create resilient humans?" who are better able to keep up the fight to address climate change. She used interviews with climate activists and extensive academic research to develop a roadmap of ten steps for climate grief (box 4.2).

THEORIES OF DENIAL

Earlier, "denial" was discussed as one of the six defense mechanisms used to avoid painful or conflicted emotions. But "denial" is a complex process with many dimensions. Sigmund Freud parsed it one way, and sociologists conceive of it another way. Denial can be accomplished in many ways—by minimizing, ignoring, rationalizing, distancing, and so on—all with the aim of trying to avoid disturbing information so we don't experience emotions of fear, mortality, guilt, and helplessness.[55] For climate change, we "know about it" in the abstract, but we can disconnect and "not know it" in our private lives and in society—its own form of denial. Two theories—supported by hundreds of studies over several decades—have great insights into the processes we use to "not know" and deny climate change.

System Justification Theory

The first theory concerns the process we use to rationalize "the way things are"—the status quo.[56] This theory says that humans tend to justify the status quo, especially in the face of a serious threat to social and economic systems (which climate change certainly presents). We evaluate existing conditions and institutions in the "system" (education, law, commerce, governance, and protection of health and environment) according to three human needs:

- *epistemic needs* (a need to maintain a sense of certainty and stability),
- *existential needs* (a need to feel safe and reassured), and
- *relational needs* (a need to affiliate with others).[57]

These needs motivate us to perceive the current system as fair, legitimate, beneficial, and stable—to think, *Oh, the system is working just fine, so we should maintain and protect it.* Any perceived threat (like climate change) to the three basic needs stimulates defensive, system-justifying responses. Rationalizing helps people cope with unwelcome realities, which happens at many levels: in beliefs, practices, social interactions, and in the system's institutions. *I don't use as much energy as other people.* Or, *If it were really a problem, the government would be doing something about it.* In many ways, it's easier to justify what's right in front of our faces – we know how it works and what to expect, and that's comfortable.

Here's an example of system justification I experienced. In the 1990s in Salt Lake City, an electric light-rail TRAX system was proposed. The timing coincided with the city hosting the 2002 Winter Olympics, for which federal transportation money was available. Opposition to the proposal was fierce: people believed it wasted money, would destroy businesses along the construction route, was dangerous, and would (tongue-in-cheek) basically destroy life as we knew it. Historical memory was absent: in the early 1900s, the city had an extensive electric trolley, and the old trolley barn is now Trolley Square shopping center.

When a greatly scaled back TRAX system (due to the opposition) opened several years later, ridership far surpassed projections and officials quickly ordered new railcars. Suburbs clamored for spur lines. Today, track miles have tripled and transit-oriented development is springing up around fifty stations. The initial proposal was hard for people to support because they could not envision it and had no prior experience with it; it was more comfortable to justify the existing system than accept a change to it. (I often wonder whether opposition would been tempered if citizens had been able to *experience* an actual light rail ride, rather than just read about it. Experiential learning is a powerful communication tool.)

Obviously, the more you benefit from the current system, the more you tend to justify it. However, even if you are being harmed by the system (that is, it's working against your self-interest, whether economic or social), you may still justify the system because it somehow seems right or patriotic to do so.[58] Thus, negative cycles and harms (such as environmental degradation) are perpetuated and real change is stifled.[59]

System justification theory recognizes that a person's reactions differ depending on whether the perceived threat is external or internal.[60] An external threat (such as nuclear posturing by another country) can create internal cohesion: the system responds and people band together against the outsider. But if the threat is internal (such as environmental problems in your state or country), a response is more difficult because something in the current system is to blame for the crisis. Facing an internal threat requires: (a) acknowledging the shortcomings in how the current system operates, (b) accepting individual and systemic responsibility for the problem, and (c) admitting that the status quo must change.[61] That's a tall order; it might seem easier to defend the system and deny the internal threat.

Psychology scholar Irina Feygina and her colleagues found in a series of experiments that system justification tendencies were associated with greater denial of ecological problems and less willingness to take pro-environmental action.[62] The greatest "system justifiers" were males, particularly those with a strong national (US) identification. One experiment found some success in reframing environmental messages to work with system justification rather than against it. Messages presented pro-environmental change as a patriotic way to preserve the "American way of life." In other words, they portrayed environmental preservation as "system-sanctioned change." (Former President Obama and the recent Green New Deal proposal also reframed "green jobs" as reducing US dependency on foreign oil and increasing security.)

Terror Management Theory

Think about all the photos and videos you see of climate-related events. The wildfire engulfing a house. A hurricane tossing boats, trees, and houses about. Crops dying on cracked and furrowed soil. People sitting on rooftops as floodwaters rise. These images powerfully prime thoughts about mortality, even if you don't consciously interpret them this way. Depending on your worldview and your identity, such messages can affect—negatively or positively—how you cope with (or deny) the threat of climate change.

Most humans (regardless of age) do not want to entertain thoughts of their mortality. When mortality thoughts do emerge, we launch an array of defenses—consciously or unconsciously—to reduce our anxiety about death. Terror Management Theory, developed by a branch of social-psychology and supported by more than 300 empirical studies over thirty years,[63] has found that mortality reminders rather predictably influence how we think and act about a variety of topics, including climate change.

Our desire for survival presents the most fundamental threat to the human self; it underpins the desire for personal control, self-worth, and social validation, and the discomfort and anxiety that may accompany those desires.[64] When mortality thoughts rise to your consciousness—*I feel a lump, That accident was a close call, My neighbor died in a flood like this*—you activate *proximal* (or conscious) *defenses* that filter disturbing information in a way that appears rational and keeps your thoughts far from death (table 4.1). We use conscious proximal defenses to deny, rationalize, and distract us from mortality reminders. For example, a friend of mine knew something was wrong in his body yet did not visit a doctor. A year later, surgery discovered that the cancer had spread and he soon died. Even though he was a nurse, his activated proximal defenses were so strong that he rationalized and denied the danger.

If you hear that the window for acting on climate change is closing, that's frightening; it's likely that your proximal defenses are activated to "protect" you from mortality thoughts. According to Cornell University emerita professor Janis Dickinson, three outcomes are likely, all of which minimize the severity of climate change, and reduce the threat.[65] One outcome is to deny that climate change exists (a *denial* that forestalls actions and solutions). Another outcome is to deny that humans are responsible for climate change

Table 4.1 Individual Psychological Defenses to Thoughts of Mortality

Proximal Defenses (conscious death thoughts)	*Distal Defenses* (unconscious death thoughts)
Denial	Transference
Distraction	Defending One's Worldview
Rationalization	Out-group Antagonism
	Self-Esteem Striving

Source: modified from Wolfe Sara E, and Amit Tubi. "Terror Management Theory and Mortality Awareness: A Missing Link in Climate Response Studies?" *Wiley Interdisciplinary Reviews: Climate Change* 10, no. 2 (2019).

(a *distraction* that "others" accountability). Finally, people may minimize the severity or project the impacts of climate change far into the future (a *rationalization* that leads to no action or small incremental responses). Even if you accept climate science, a primed mortality thought can quickly produce a proximal defense that emotionally distances the threat.

It is possible, however, for a person's terror management to result in positive outcomes, based on worldview, identity, and the norms of behavior of the individual and her/his social groups. If you (and your social groups) have an altruistic worldview, a threat about climate change could prompt you to become involved in groups and communities that are active in environmental issues and climate change solutions.[66]

As mortality reminders increase or worsen—perhaps through steady news of sea level rises and record temperatures—the person may divert mortality thoughts into his unconscious and not actively think about them. Then, proximal defenses switch to *distal defenses*, of which there are four kinds.

First, *transference* directs attention elsewhere, such as toward charismatic figures (with similar worldviews), or through blind following and reduced rational criticism of public figures (such as politicians or celebrities).[67] One study found that climate change threats increased preferences for authoritarian leadership and distancing of out-groups (such as immigrants).[68]

Defending one's worldview—including bolstering or clinging to it fervently—is a distal defense. The worldview might be anthropocentrism (belief in human domination and control of nature) or a religious belief that puts everything in God's hands and free of human responsibility. Dickinson called the chanting of "drill, baby, drill" at the 2008 Republican Convention in Minneapolis an example of fervently defending a worldview.[69]

Showing *antagonism toward out-groups* is a defense when one feels (unconsciously) threatened. This happens between those in different religions, political parties, or between environmentalists and anti-environmentalists. This distal defense is worrisome. Dickinson says the clash between disparate ideologies is likely to produce even deeper rifts around the world, and it can involve devaluation, marginalization, aggression, and even violence against individuals and groups who hold opposing worldviews.

Finally, *striving for self-esteem* is a defense we use to feel better when anxiety related to climate change unconsciously increases. This may be acted out with increased status-driven consumerism and materialism,[70] which of course *increases* the threat of climate change by increasing carbon emissions. This distal defense is used by people with greatly differing worldviews; it's possible to encounter both a "materialistic environmentalist" as well as someone with a conservative sense of entitlement who both increase consumption. Studies have found that death reminders reliably increase humans' selfish exploitation of natural resources.[71]

On the surface, you might not connect materialism or antagonism to climate anxiety because these defenses are *unconscious*. The defenses do not reduce the threat and can be counterintuitive or counterproductive because individuals will incur greater rather than lesser danger by avoiding and postponing action on climate change. But it's one way our internal psychology processes threats to mortality.

Examples of these defenses are present every day in news stories, where narratives are skewed toward negative "disaster and death" visions. These visions, unsurprisingly, may exacerbate counterproductive responses to the climate crisis because they trigger defenses to reduce our anxiety about mortality. One way to counteract the emotions surrounding mortality is to accept that everything you love (and have), you will eventually lose[72]—a recognition of impermanence and mortality of all life, which is a truth that most humans struggle to accept.

A communication lesson from Terror Management Theory is that messages need to integrate the mortality threat with an effective response. One study effectively coupled threat messages with things that could be done (efficacy options) to reduce the threat.[73]

The distal defense of striving for self-esteem also has potential for positive action in the form of immortality or "hero projects," which we'll return to in a moment. Another strategy to counter terror management tendencies is to focus energy and narratives on positive visions of a future, albeit greatly changed, world. Disaster-and-death sells better on the news, but who wouldn't like to live in a fossil-fuel-free world if it meant cleaner air and water, quieter cities, and simpler, healthier lives and communities?

EMOTIONAL RESILIENCE

In her novel *Flight Behavior*, Barbara Kingsolver's protagonist lives a hardscrabble life in rural Tennessee.[74] For an inexplicable reason, one winter day millions of monarch butterflies come to roost on the mountain next to Dellarobia Turnbow's house. Biologists park their trailer near her barn to study these (fictional but somewhat realistic) climate change refugees. As Dellarobia learns more about climate change, she declares, "There's no room for the end of the world at my house." Her defenses are up and her first instinct is to deny the terror and reality of the climate crisis. However, on the Six Americas spectrum presented in chapter 1 (alarmed, concerned, cautious, disengaged, doubtful, dismissive), she was "doubtful" and open to learning more.

Dellarobia's *ontological security* was threatened by climate change, or "the confidence that most human beings have in the continuity of their self-identity and the constancy of the surrounding social and material environments."[75]

Climate change is already creating a deep shift in our cultural identities—the way we view ourselves and who we are in our social networks and the larger society—and how we value (and haven't valued) the living world and our place in it.

Dellarobia exhibited the qualities of *emotional resilience*, which means she was able to adapt to a stressful situation like the climate crisis when she became aware of it. The Latin word "resilio" means to bounce back, so emotionally that's an ability to feel your feelings and acknowledge them, but not let them paralyze or derail you. Emotionally resilient people are aware of their emotions, they persevere when facing difficulty, and they seek social support when needed. Other helpful characteristics for resiliency are optimism and a sense of humor.

The American Psychological Association report mentioned above provides tips to support good mental health in a changing climate.[76] First, build a belief in your own resilience and foster optimism. Also, cultivate skills for coping with and self-regulating your emotions. Keep doing the things that give you a sense of meaning. And, engage in things that connect you to family, place, culture, and community—and I would add, Eairth.

Developing emotional resiliency is a unique process for each person. It can help to first understand your emotions (box 4.1) and grief (box 4.2). The next sections discuss additional ways to develop emotional resiliency: accepting a changed "normal" world, boosting self-esteem through hero projects, and seeking emotional support in fellowship with others.

Accepting a Changed "Normal"

That's a big ask—for climate change, for life under a pandemic, for civil unrest, for many things. The only truly "normal" part of life is that it changes constantly, but we still get attached to what we perceive as normal (even if it's abnormal and unjust). Accepting a changed climate doesn't mean you don't also mourn and lament (and are angry) that it happened. Accepting present reality requires learning how to "let go of gone things"[77] which is necessary in order to create new realities, such as a planet where all flourishing is mutual.

I believe that much of the anomie, angst, and instability playing out in so many places in the world today is the manifestation of (and coming to terms with) the new inconstancy of constant nature. Throughout human history, we believed Eairth would remain more or less changeless, no matter what we did.

Acclaimed Welsh author Jay Griffiths says that for thousands of years, humanity believed in the relative changelessness of the natural world. Other than seasons, migrations, and an occasional flood, nature stayed the same, and its gradual, slow change was our psychological reassurance: "flux within fixity, mutability with larger immutability, unpredictable weather with predictable climates. Climate change does, indeed, go against the grain of thousands of years of accumulated belief that the climate and larger environment is stable and must be so: if it is not, the easefulness and stability of the human mind is threatened."[78]

Griffiths argues that we now increasingly seek constancy not from the natural world but in the so-called built environment and the "pseudo-climate of the economic system," which is given the attention and concern that the real climate needs. "Financial well-being and economic stability are prized more than ecological well-being and climate stability."[79] Nevertheless, climate disruptions are changing both.

The US National Climate Data Center released "new normals," or averages for temperatures, rainfall, and snow in particular regions from 1980 to 2010. It was no surprise that the numbers in this thirty-year period had changed and warmed in most all regions. These new "normals" essentially codify and camouflage human-driven warming. When the weather report says today's temperature was "normal" for that day, it disguises the fact that your day's temperature was likely not "normal" but above it. Celebrating the "normal" fits nicely with system justification and terror management theories; everything is fine.

Yet, "normal" is already so altered that an atmospheric chemist proposed renaming the current geologic epoch the "Anthropocene."[80] Human actions have undoubtedly altered climatic, biophysical, and evolutionary processes on a planetary scale, but in many ways the label is more about politics than geology; philosopher Carlos Santana constructs a detailed argument that the geologic record cannot truly distinguish the Anthropocene from the current Holocene.[81]

Some argue that the proposed name-change would emphasize the unpredictability and chaos that human-caused climate change brings to the planet, while others find the proposed name full of anthropocentric hubris. Feminist theorist Donna Haraway bristles at Anthropocene discourse, calling it "simply wrong-headed and wrong-hearted."[82] Instead, she offers provocative new ways to reconfigure our relations to the earth and all its inhabitants, where the human and non-human are inextricably linked in "tentacular" practices more conducive in building more livable futures. Environmental philosophers have other conceptualizations of the epoch. Glen Albrecht proposes the "Symbiocene" after "symbiosis" or living together for mutual benefit,[83] while David Abram introduces the Humilocene, a grounded-human epoch steeped in both humiliation and humility.[84]

As someone once said: it's not that we do not want to change, but that we do not want to *be* changed. If we get wind that the future may not look like the present (and it rarely does), we feel threatened. We cannot fathom—or even envision—the changes to come, in our own lives and for Eairth. All we know is that it must surely mean "sacrifice." Biologist and writer Carl Safina questions this self-centered notion of sacrifice:

> As though war built around oil is not sacrifice. As though losing polar bears, ice-dependent penguins, coral reefs, and thousands of other living companions is not sacrifice. As though withered cropland is not a sacrifice, or letting the fresh water of cities dry up as glacier-fed rivers shrink. As though risking seawater inundation and the displacement of hundreds of millions of coastal people is not a sacrifice—and reckless risk. *But don't tell me to own a more efficient car; that would be a sacrifice!* We think we don't want to sacrifice, but sacrifice is exactly what we're doing by perpetuating problems that only get worse; we're sacrificing our money, and sacrificing what is big and permanent, to prolong what is small, temporary, and harmful.[85]

Climate scientist and communicator Katharine Hayhoe believes that our tendency to focus on what we might have to "give up" is due to a "solution aversion" where we fear solutions more than we fear climate change.[86] Hayhoe says when people believe that a price on carbon "will destroy the economy" and "I won't be able to drive my truck anymore," what's really missing is a vision of a *better* future. She says that climate change tends to confront us with two apocalyptic scenarios—one, the end of human civilization, and two, that we will have to throw away everything that makes life comfortable—and more people are afraid of the latter. Instead, she believes we need to talk about and envision a new normal—a future with sufficient energy, water, and abundant food—and then choose solutions that move us in that direction.

By the end of *Flight Behavior*, Dellarobia knew a lot about climate change and what it meant, particularly for her young children's world. She had witnessed significant change on her home ground (a farm) and disruptions to seasons and climate that greatly impacted jobs and livelihoods. Grasping this immense problem put her own life and relationships into sharper focus; there were bigger, more pressing problems in the world than her own.

Hero Projects

A second path to emotional resiliency involves reframing "mortality" in a way that positively boosts self-esteem. If Terror Management Theory predicts a reaction to our fears of mortality, might that fear be reduced by focusing instead on our *immortality*? Janis Dickinson says yes and believes that "immortality projects"—also called "hero projects"—can redirect fears into positive energy and solutions for climate change.[87]

Decades ago, cultural anthropologist Ernest Becker held that the psychological basis of emotional and mental health was holding tightly to anxiety-buffering "immortality projects" that promoted self-esteem and meaning.[88] We seek social recognition and a sense of importance that will extend beyond our biological existence. Although immortality is equated with money and material acquisition, these objects lack heroism and hence come up empty or inspire guilt.

Instead, Dickinson suggests finding self-esteem and heroism in service of the natural world. She reminds us that President Franklin Roosevelt's Civilian Conservation Corp was a way for individuals to better themselves and repair environmental degradation. Historically, nature has played an integral part in immortality-striving rituals and symbolism and provides a good context for self-esteem, heroism, and spirituality.

Dickinson suggests hero projects that leave a lasting legacy (and hence "immortality") involving birds. Bird populations have plummeted due to climate change and habitat degradation, so an individual's hero project could engage social actors and groups to preserve bird migration corridors, restore habitat from climate impacts, or introduce all teens in a state to bird appreciation. Collective work and social support would increase emotional resiliency for all.

Although hero projects could involve any climate-related topics, Dickinson chose birds because they take to the sky with beauty, mystery, and charisma, which makes them an ideal "transference object" where individuals can project their striving for immortality. These immortality totems connect vitality with flight. She cites studies that have found that idealizing birds can have anxiety-buffering effects.[89] In one study, participants primed with death thoughts were more likely to express the desire to fly. In a second study, when participants visualized a flight-fantasy, it ameliorated terror. Dickinson concludes that birding can foster resilience when confronting climate change.

Bird-watchers form a "culture of honor" that provides self-esteem, social interaction, and even competition where heroes emerge. Programs that reestablish connections with nature through birds are the Cornell Laboratory of Ornithology's Celebrate Urban Birds, and the environmental education program Flying Wild. Youth from all backgrounds learn about birds, reduce fears of nature, and get important unstructured time outside.

Dickinson concludes that hero projects have great promise for alleviating some of the terror of climate change: "If true change requires both heroic leadership and a cultural context for the heroism of many, a cultural worldview that incorporates both innovation and idealization of the natural world is the logical immortality project and the best opportunity for heroism in these times. Love of nature is a deep ethical and spiritual issue that is consistent with most belief systems . . . [and] provides profound opportunities for symbolic immortality."[90]

Fellowship

A final route to emotional resilience is facing the climate crisis in fellowship with others. The crisis is global and systemic and no one person acting alone—no matter how powerful—can address it sufficiently or navigate its complex emotions.

One of the hardest parts about climate change is not just the huge threat and long odds of addressing it, but the idea of facing all that alone. As journalist David Roberts notes, what people are groping for is *fellowship*; if people feel part of a community dedicated to a common purpose, people can face overwhelming odds with good spirits.[91] Fellowship is where people find optimism and grow self-efficacy and self-esteem.

You can find emotional fellowship with people with whom you feel safe, and among people with whom you share goals or values—friends, family, colleagues, or groups involved in climate change. Support that helps buffer anxiety may come in the form of conversations and sharing, and also from active "doing" such as volunteer work or outdoor activities. Along the way, fellowship will soon flow naturally.

Finally, for emotional resiliency, turn and go outside. Go into the woods, or a prairie, the desert, even a city park or a cornfield. Smell it, feel it, touch it, and above all breathe it. Getting acquainted and grounded with the living world is a great way to recharge drained reserves and calm fears. We all need joyful reminders that Eairth is miraculous and worthy of care.

I went outside to learn about grief and emotional resilience after the wildfire. The first summer, I was grief-stricken and surrounded by black char; my beloved woods were gone. The land would never be what it once was; some western forests will not grow back due to climate changes. But by the second summer, I was learning to "let go of gone things." The resiliency of animals, birds, wildflowers, shrubs, and a dozen half-burned trees astounded me. Their tenacity, and the support of neighbors and friends, helped me tremendously to navigate the emotional terrain of climate change.

NOTES

1. Matthew Taylor and Jessica Murray, "'Overwhelming and Terrifying': The Rise of Climate Anxiety," *The Guardian*, February 10, 2020, https://www.theguardian.com/environment/2020/feb/10/overwhelming-and-terrifying-impact-of-climate-crisis-on-mental-health/.

2. Scott Nadler, "How to Reject 'Climate Porn' and Reach 'Climate Acceptance,'" *GreenBiz*, October 2, 2018, Accessed December 18, 2018, https://www.greenbiz.com/article/how-reject-climate-porn-and-reach-climate-acceptance/.

3. Renee Lertzman, *Environmental Melancholia: Psychoanalytic Dimensions of Engagement* (New York, NY: Routledge, 2015).

4. Lawrence A. Palinkas, "One of the Most Overlooked Consequences of Climate Change? Our Mental Health," *The Daily Climate*, December 2, 2019, https://www.dailyclimate.org/how-climate-change-affects-mental-health-2641491618.html.

5. "Mental Health and Our Changing Climate: Impacts, Implications, and Guidance," *American Psychological Association and ecoAmerica*, March 2017, p. 3.

6. S. R. Lowe, E. E. Manove, and J. E. Rhodes, "Posttraumatic Stress and Posttraumatic Growth among Low-Income Mothers Who Survived Hurricane Katrina," *Journal of Consulting and Clinical Psychology* 81, no. 5 (2013): 877–89.

7. Bob Carroll, Hazel Morbey, Ruth Balogh, and Gonzalo Araoz, "Flooded Homes, Broken Bonds, the Meaning of Home, Psychological Processes and Their Impact on Psychological Health in a Disaster," *JHAP Health and Place* 15, no. 2 (2009): 540–47.

8. Richard A. Bryant, Lisa Gibbs, Hugh Colin Gallagher, et al., "Longitudinal Study of Changing Psychological Outcomes Following the Victorian Black Saturday Bushfires," *Australian and New Zealand Journal of Psychiatry*, https://nls.ldls.org.uk/welcome.html?lsidyva32153f7.

9. Iris-Tatjana Kolassa, Verena Ertl, Cindy Eckart, Stephan Kolassa, Lamaro P. Onyut, and Thomas Elbert, "Spontaneous Remission from PTSD Depends on the Number of Traumatic Event Types Experienced," *Psychological Trauma: Theory, Research, Practice* 2, no. 3 (2010): 169–74.

10. Somini Sengupta, "Extreme Weather Displaced a Record 7 Million in First Half of 2019," *New York Times*, https://www.nytimes.com/2019/09/12/climate/extreme-weather-displacement.html?rref=collection%2Fsectioncollection%2Fclimate&action=click&contentCollection=climate&region=stream&module=stream_unit&version=latest&contentPlacement=1&pgtype=sectionfront.

11. Ashlee Cunsolo and Neville R. Ellis, "Ecological Grief as a Mental Health Response to Climate Change-Related Loss," *Nature Climate Change* 8 (April 2018): 275–81, 275.

12. "Mental Health and Our Changing Climate," 25.

13. John Simister and Cary Cooper, "Thermal Stress in the U.S.A.: Effects on Violence and on Employee Behaviour," *Stress and Health* 21, no. 1 (2005): 3–15.

14. Jennifer A. Fresque-Baxter and Derek Armitage, "Place Identity and Climate Change Adaptation: A Synthesis and Framework for Understanding," WCC *Wiley Interdisciplinary Reviews: Climate Change* 3, no. 3 (2012): 251–66; Pia Schönfeld, Julia Brailovskaia, Angela Bieda, Xiao Chi Zhang, and Jürgen Margraf, "The Effects of Daily Stress on Positive and Negative Mental Health: Mediation through Self-Efficacy," *International Journal of Clinical and Health Psychology* 16, no. 1 (2016): 1–10.

15. "Mental Health and Our Changing Climate," 25.

16. A. Cunsolo Willox, S. L. Harper, J. D. Ford, V. L. Edge, K. Landman, K. Houle, S. Blake, and C. Wolfrey, "Climate Change and Mental Health: An Exploratory Case Study from Rigolet, Nunatsiavut, Canada," *Climatic Change* 121,

no. 2 (2013): 255–70; Neville Ellis and Glenn Albrecht, "Climate Change Threats to Family Farmers' Sense of Place and Mental Wellbeing: A Case Study from the Western Australian Wheatbelt," *Social Science & Medicine* 175 (2017): 161–68.

17. David Corn, "It's the End of the World as They Know It: The Distinct Burden of Being a Climate Scientist," *Mother Jones*, July 8, 2019.

18. Lesley Head and Theresa Harada, "Keeping the Heart a Long Way from the Brain: The Emotional Labour of Climate Scientists," *Emotion, Space and Society* 24 (2017): 34–41.

19. Heidi Hendersson and Christine Wamsler, "New Stories for a More Conscious, Sustainable Society: Claiming Authorship of the Climate Story," *Climatic Change*, November 28, 2019, doi:10.1007/s10584-019-02599-z.

20. *America's Top Fears 2018, Survey of American Fears*, Chapman University, https://blogs.chapman.edu/wilkinson/2018/10/16/americas-top-fears-2018/.

21. Gallup lumps "climate change" with "environment" and "pollution," which in 2019 were mentioned as the most important issues by just 4 percent of respondents. See www.gallup.com.

22. Palinkas, "One of the Most Overlooked Consequences of Climate Change?"

23. S. C. Moser and L. Dilling, "Making Climate Hot: Communicating the Urgency and Challenge of Global Climate Change," *Environment* 46, no. 10 (2004): 32–46; S. O'Neill and S. Nicholson-Cole, "Fear Won't Do It: Promoting Positive Engagement with Climate Change Through Visual and Iconic Representations," *Science Communication* 30, no. 3 (2009): 355–79; M. Chen, "Impact of Fear Appeals on Pro-environmental Behavior and Crucial Determinants," *International Journal of Advertising* 35, no. 1 (2015): 1–19.

24. S. C. Moser, "More Bad News: The Risk of Neglecting Emotional Responses to Climate Change Information," in *Creating a Climate for Change*, ed. S. C. Moser and L. Dilling (Cambridge: Cambridge University Press, 2007), 64–80.

25. Christiane Amanpour, "Climate Scientist: Think with Heart, Not Head," *CNN*, January 8, 2019, https://www.cnn.com/videos/world/2019/01/08/katharine-hayhoe-amanpour-climate-change.cnn/.

26. Sabine Roeser, "Risk Communication, Public Engagement, and Climate Change: A Role for Emotions," *Risk Analysis* 32, no. 6 (2012): 1033–40.

27. Antonio Damasio, *Descartes' Error* (New York, NY: Putnam, 1994).

28. Paul Slovic, "Numbed by Numbers," *Foreign Policy*, 2007.

29. Renee Lertzman, "Breaking the Climate Fear Taboo," *Sightline*, March 2, 2014, https://www.sightline.org/2014/03/12/breaking-the-climate-fear-taboo/.

30. Lertzman, *Environmental Melancholia*.

31. Kristin Haltinner and Dilshani Sarathchandra, "Climate Change Skepticism as a Psychological Coping Strategy," *Sociology Compass* 12, no. 6 (2018).

32. *National Climate Assessment 2018*, https://nca2018.globalchange.gov/.

33. Lertzman, *Environmental Melancholia*, 105.

34. Cunsolo and Ellis, "Ecological Grief."

35. Lertzman, *Environmental Melancholia*, 6.

36. Howard Searles, *The Nonhuman Environment in Normal Development and in Schizophrenia* (New York, NY: International Universities Press, 1960).

37. A. Cunsolo and K. Landman, eds., *Mourning Nature: Hope at the Hearth of Ecological Loss and Grief* (Montreal: Queen's University Press, 2017), 10.

38. Lertzman, *Environmental Melancholia*; Cunsolo and Ellis, "Ecological Grief."

39. Lertzman, "Breaking the Fear Taboo."

40. Renee Lertzman, *Permission to Care*, https://reneelertzman.com/permission-to-care/.

41. Eleanor Busby, "Climate Change Activism 'Reducing Mental Health Symptoms among Young People'," *The Independent*, November 28, 2019, https://www.independent.co.uk/news/education/education-news/climate-change-school-strikes-mental-health-students-children-a9225081.html.

42. Lertzman, *Environmental Melancholia*.

43. Rosemary Randall, "Loss and Climate Change: The Cost of Parallel Narratives," *Ecopsychology* 1, no. 3 (2009): 118–28.

44. Cunsolo and Landman, *Mourning Nature*.

45. Cunsolo and Ellis, "Ecological Grief."

46. Ashlee Cunsolo Willox, "Climate Change as the Work of Mourning," *Ethics & the Environment* 17, no. 2 (2012): 145.

47. Cunsolo and Ellis, "Ecological Grief."

48. Judith Butler, *Precarious Life: The Power of Mourning and Violence* (New York, NY: Verso, 2004).

49. Cunsolo Willox, "Climate Change as the Work of Mourning," 137–64, 151.

50. E. Kübler-Ross and D. Kessler, *On Grief and Grieving: Finding the Meaning of Grief Through the Five Stages of Loss* (New York, NY: Scribner, 2007).

51. Nadler, "How to Reject 'Climate Porn'."

52. William Worden, *Grief Counseling and Grief Therapy* (London: Tavistock, 1983).

53. Robert Neimeyer and Joanne Cacciatore, "Toward a Developmental Theory of Grief," in *Techniques of Grief Therapy: Assessment and Intervention*, ed. R. Neimer (New York, NY: Routledge, 2016), 3–13.

54. Daisy Simmons, "Sad About Climate Change? There's a Support Group for That," *Yale Climate Connections*, November 28, 2016.

55. Kari Marie Norgaard, "Climate Denial: Emotion, Psychology, Culture, and Political Economy," in *The Oxford Handbook of Climate Change and Society*, ed. J. S. Dryzek, R. B. Norgaard, and D. Schlosberg (New York, NY: Oxford University Press, 2011).

56. Irina Feygina, John T. Jost, and Rachel E. Goldsmith, "System Justification, the Denial of Global Warming, and the Possibility of 'System-Sanctioned Change'," *Personality and Social Psychology Bulletin* 36, no. 3 (2010): 326–38; J. T. Jost, I. Liviatan, J. van der Toorn, A. Ledgerwood, A. Mandisodza, and B. A. Nosek, "System Justification: How Do We Know It's Motivated?," in *The Psychology of Justice and Legitimacy: The Ontario Symposium*, Vol. 11, ed. R. Bobocel et al. (Hillsdale, NJ: Lawrence Erlbaum, 2009).

57. J. T. Jost, A. Ledgerwood, and C. D. Hardin, "Shared Reality, System Justification, and the Relational Basis of Ideological Beliefs," *Social and Personality Psychology Compass* 2 (2008): 171–86.

58. P. J. Henry and A. Saul, "The Development of System Justification in the Developing World," *Social Justice Research* 19 (2006): 365–78.

59. C. Wakslak, J. T. Jost, T. R. Tyler, and E. Chen, "Moral Outrage Mediates the Dampening Effect of System Justification on Support for Redistributive Social Policies," *Psychological Science* 18 (2007): 267–74.

60. J. Ullrich and C. Cohrs, "Terrorism Salience Increases System Justification: Experimental Evidence," *Social Justice Research* 20 (2007): 117–39.

61. Feygina et al., "System Justification."

62. Feygina et al., "System Justification."

63. T. Pyszczynski, J. Greenberg, S. Solomon, and M. Maxfield, "On the Unique Psychological Import of the Human Awareness of Mortality: Theme and Variations," *Psychological Inquiry* 17 (2006): 328–56; Sara E. Wolfe and Amit Tubi, "Terror Management Theory and Mortality Awareness: A Missing Link in Climate Response Studies?," *Wiley Interdisciplinary Reviews: Climate Change* 10, no. 2 (2019).

64. J. Juhl and C. Routledge, "Putting the Terror in Terror Management Theory: Evidence That the Awareness of Death Does Cause Anxiety and Undermine Psychological Well-Being," *Current Directions in Psychological Science* 25, no. 2 (2016): 99–103, doi:10.1177/0963721415625218/.

65. Janis Dickinson, "The People Paradox: Self-Esteem Striving, Immortality Ideologies, and Human Response to Climate Change," *Ecology and Society* 14, no. 1 (2009): 34.

66. K. E. Vail and J. Juhl, "An Appreciative View of the Brighter Side of Terror Management Processes," *Social Sciences* 41020–45 (2015), doi:10.3390/socsci4041020/.

67. Dickinson, "The People Paradox."

68. I. Fritsche, E. Jonas, D. N. Kayser, and N. Koranyi, "The Malicious Effects of Existential Threat on Motivation to Protect the Natural Environment and the Role of Environmental Identity as a Moderator," *Journal of Environmental Psychology* 30, no. 1 (2010): 67–79, doi:10.1177/0013916510397759/.

69. Dickinson, "The People Paradox."

70. T. Kasser and K. M. Sheldon, "Of Wealth and Death: Materialism, Mortality Salience, and Consumption Behavior," *Psychological Science* 11 (2000): 348–51.

71. K. M. Sheldon and H. A. McGregor, "Extrinsic Value Orientation and 'The Tragedy of the Commons'," *Journal of Personality* 68, no. 2 (2000): 383–411, doi:10.1111/1467-6494.0010/.

72. Francis Weller, *The Wild Edge of Sorrow: Rituals of Renewal and the Sacred Work of Grief* (Berkeley, CA: North Atlantic Books, 2015).

73. W. Xue, D. W. Hine, A. D. G. Marks, W. J. Phillips, P. Nunn, and S. Zhao, "Combining Threat and Efficacy Messaging to Increase Public Engagement with Climate Change in Beijing, China," *Climatic Change* 137, no. 1 (2016): 43–55, doi:10.1007/s10584-016-1678-1/.

74. Barbara Kingsolver, *Flight Behavior* (New York, NY: Harper, 2012).

75. A. Giddens, *The Consequences of Modernity* (Stanford, CA: Stanford University Press, 1990).

76. "Mental Health and Our Changing Climate," 7.
77. Upile Chisala, *Nectar* (Kansas City, MO: Andrews McMeel Publishing, 2019).
78. Jay Griffiths, "Myths of Stability: Putting Capitalism Before Creation," *Orion*, November–December 2013, pp. 13–14.
79. Griffiths, "Myths of Stability."
80. P. J. Crutzen and E. F. Stoermer, "The 'Anthropocene'," *International Geosphere-Biosphere Program Newsletter* 41 (2000): 17–18.
81. Carlos Santana, "Waiting for the Anthropocene," *The British Journal for the Philosophy of Science* 70, no. 4 (2019): 1073–96.
82. Donna J. Haraway, *Staying with the Trouble: Making Kin in the Chthulucene* (Durham, NC: Duke University Press, 2016), 50.
83. Glenn A. Albrecht, "Exiting the Anthropocene and Entering the Symbiocene," December 17, 2015, https://glennaalbrecht.wordpress.com/2015/12/17/exiting-the-anthropocene-and-entering-the-symbiocene-via-sumbiocracy-symbiomimicry-and-sumbiophilia/.
84. David Abram, "Interbreathing Ecocultural Identity in the Humilocene," in *Routledge Handbook of Ecocultural Identity*, ed. Tema Milsein and Jose Castro-Sotomayor (Boca Raton, FL: Routledge, 2020).
85. Carl Safina, "The Moral Climate," *Orion*, September–October 2008, pp. 58–59.
86. Amanpour, "Climate Scientist."
87. Dickinson, "The People Paradox."
88. Ernest Becker, *The Denial of Death* (New York, NY: Free Press, 1973).
89. S. Solomon, J. Greenberg, T. Pyszczynski, F. Cohen, and D. Ogilvie, "Teach These Souls to Fly: Supernatural as Human Adaptation," in *Evolution, Culture, and the Human Mind*, ed. M. Schaller, A. Norenzayan, S. Heine, T. Yamagishi, and T. Kameda (Mahwah, NJ: Lawrence Erlbaum, 2010), 99–118.
90. Dickinson, "The People Paradox," 11.
91. David Roberts, "The Case for 'Conditional Optimism' on Climate Change," *Vox*, December 31, 2018, https://www.vox.com/energy-and-environment/2018/12/28/18156094/conditional-optimism-climate-change.

Chapter 5

Breaking the Silence

Strategies for Talking about Climate Change

During a New Year's Day hike on a mesa in southern Utah, my friend's friend asked me about the book I was writing. When he learned the topic was climate change, he quipped, "Yeah, well, Al Gore keeps jet-setting around the world." He was flight-shaming Gore for his consumer behavior as though it were hypocritical for a climate change activist to use fossil fuels this way.[1]

His quip provoked the conversation we all dread. We avoid bringing up the C-words in polite company or with strangers. We fear the remark that blind-sides, the loaded come-back, the denier-speak, or the selective science-'splaining, so we keep our tongues in our mouths. And, it's created overwhelming *climate silence*. It's the topic that cannot be named, let alone discussed.

I said to him, "Well, you and I fly, too, so I guess we're all implicated and responsible." We kept talking as we climbed the sandy wash. Eventually, we agreed that climate change was real and we were both concerned about it, even though we disagreed on some of the particulars.

At some point, his daughter entered the conversation with a series of retorts from the climate denier playbook: the global warming hoax was destroying coal jobs and the economy. Ah, she was in the 10 percent "dismissive" camp (from the Six Americas study[2]) and it wasn't worth trying to have a conversation with her. But her father, despite his Al Gore remark, was somewhere in the other 90 percent; I was glad we talked and found some common ground.

Since then, I have learned a lot about breaking the silence on climate change—how to broach the subject, what to say, how to listen, what to avoid, and (importantly) when to disengage. My conversation goal is never to change someone's mind, but to bring climate change into the light where it belongs, to give it some air. And no matter the outcome, breaking the silence about climate change is one of the most important social things you can do to help. Talking can bring you into fellowship with others, remove "shame" from living in fossil fuel culture, and reinforce the urgency of moving forward together. The climate crisis needs to be on the lips of the broadest, most cross-cutting group of people possible: young and old, every color and every faith, tree-huggers and tree-cutters, from every point on the political spectrum.

The new direction presented in this chapter is encouraging and preparing individuals to talk about climate change—not just to like-minded peers but to family, neighbors, colleagues, and strangers. Effective interpersonal conversation about climate change is a foundational skill for other strategies, such as engaging in collective action and communicating with institutions and leaders. I'm not saying talking is easy, but in my experience (and that of many students), talking about climate change *works*. Talking can elevate the importance of the climate crisis throughout culture and break social inertia. The chapter explores the aims of conversation, strategies and entry points, and what to do when the talking gets tough. Mock conversations with "friendly skeptics" and role-playing exercises prepare you for climate conversations.

TALKING WORKS

Fifty-nine percent of Americans say they "rarely" or "never" discuss climate change with family and friends. Fifty-six percent of Americans say they hear about it in the media at least once a month (although it's present far more frequently). But at the same time, 66 percent say they are "somewhat" or "very" worried about it.[3]

If you rarely talk about something you're worried about, you are stuffing the worry deep inside. Maybe you think you don't know enough science to talk about it, or you're afraid your conversation will start a fight. You might feel normative pressure to clam up; in your social networks the topic might be seen as controversial, so you self-censor. And of course, climate change is now a highly polarized issue in the US (even though thermometers have no political affiliation[4]). Those reasons sound logical, but silence is not a productive way to deal with something that most people consider important and already worry about.

From a decade of my climate conversations (and those of my students), I know talking produces positive outcomes. Peer-to-peer conversations can lead to a deeper, more sophisticated understanding of complex issues such as climate change.[5] Talking can ameliorate worry and create rippling effects among those you talk to. Some people will be eager to finally talk about it. And, you might learn that you greatly underestimate how many people think climate change is real (which can affect your own beliefs). One survey found that people estimate that about half of other Americans think global warming is happening, when in fact nearly 70 percent do.[6]

Talking with others can create important shifts and effects, particularly among those closest to you. I've heard repeatedly that some students had never talked about climate change with partners, parents, or close friends until a class assignment prompted them to. Just one conversation can open a door—to additional talks and questions, or letting conservative grandpa know that the climate crisis is very important to his granddaughter.

As a social actor, your beliefs are very influential to the beliefs of those closest to you. The perceived degree of agreement about climate change is important to the climate change beliefs and attitudes for people across the ideological spectrum, but especially among conservatives.[7] Because most conservatives identify more strongly with friends and family than conservatives as a broad category, if their own friends and family care about climate change, they are more likely to care as well. If people mistakenly believe that others in their lives do not care about climate change, it's important to encourage them (especially conservatives) to talk about climate change. Researchers recommend that communicators try to engage people based on their non-political identities (such as outdoor activities or grandparenting), which may enable more constructive conversations about climate change than partisan discourse.[8]

It's helpful to think of talking about climate change not as single conversation but as a process or even a "loop." One study found that climate conversations with friends and family enter participants into a "pro-climate social feedback loop."[9] When discussing climate change with friends and family, it's likely that some people will learn influential facts, such as the scientific consensus that it's happening and is human-caused. The feedback loop occurs when learning encourages further discussion and deeper engagement. In conversations with close friends and family, people less engaged with the issue may be more receptive than when an identical message is communicated by someone not part of their close social network.

Kids talking with their parents about climate change can also be very influential.[10] In one study, ten-to-fourteen-year-olds completed a school curriculum about climate change that included an outdoor science project (like collecting plankton), blog posts about what they learned, and interviews of their parents about changes they had witnessed in their local climate. The parents most influenced by these conversations were conservative men, and girls were more influential than boys on their parents' views. Parents generally trust what their children tell them, so kids sharing what they had been learning increased parental concern.

Although many individuals "inherit" their climate change beliefs (at least in part) from their family of origin, it's plausible that talking and learning more could shift these views. Consider this person: *I really doubted climate change for a while, because honestly it scared me. I figured if I just denied it and pretended it wasn't a thing, it wouldn't be and it would just go away.*[11]

She switched from a climate denier to an acceptor and recognized that the basis of her denial was fear. She responded to an online survey: "Former climate deniers, what changed your mind?" This (unscientific) survey found that the most common origin of respondents' skeptical beliefs was their home environment: family and partisan values. For some, what changed their minds was learning about the science and deciding that scientists were more trustworthy than deniers. Others said it was because humans should take care of the planet. For others, it was talking with people and firsthand experience with warm and weird weather.

A scientific study also noted attitude changes about climate change. Over a three-year period, about 8 percent of Americans shifted their concern; of those who changed their concern in a clear direction, 84 percent of them became more concerned (almost equal proportions of Democrats and Republicans).[12] Of those whose concern increased, they reported experiencing or hearing about climate impacts, realizing its seriousness, and becoming more informed.

One group experiencing impacts and talking more about climate change is farmers. In the US farm belt, 2019 was a terrible year with historic rain, disastrous flooding, and a record 20 million acres that couldn't be planted. Iowa farmer (and political conservative) Ray Gaesser said, "Everybody I talk to, including farmers, they say 'yeah we need to talk about this.' We need to find ways to adapt to what's going on. We're seeing things we're not used to seeing."[13] Increasingly, ag sectors and state farm bureaus convene discussions for farmers, such as how to reorient farms to sequester carbon and rebuild soil (called regenerative agriculture). Interfaith Power & Light has pulled farmers into small groups to talk about climate change in religious settings. The non-profit Solutions from the Land hosted farmer-led discussions on climate change all over the country. The discussions help farmers learn how to become more resilient and that their experience with unpredictable, extreme weather is similar to their neighbors.

It's easy to let our anxieties about the climate crisis turn us inward where our fears lead to paralysis, silence, and inaction. Talking with those closest to you is a positive place to start. As scientist and climate communicator Katharine Hayhoe says, "If we don't talk about climate change, why would we care? If we don't care about it, why would we act? So, action begins with a conversation."[14]

WHY IT'S SO HARD TO TALK ABOUT CLIMATE CHANGE

My pep talk is not to say that talking about the climate crisis is easy. It's an extremely complex problem that triggers deep emotions, many of which live silently at the unconscious level. It's important to know a bit about why it's hard to talk about climate change so that you head into conversations with eyes wide open.

Public opinion surveys report that most people are aware of climate change (or global warming), are familiar with the terms, and 70 percent of Americans think it's happening.[15] The high level of public awareness, however, does not mean it's something we truly grasp on a deeper level. Philosopher Robert Kirkman contends that the threat of climate change is something we know *about*, but it is not something that we always *feel*, so it remains rather *theoretical* and hard to grasp as a danger.[16] The risk feels nonpersonal and seems to involve the future, other places, other people, other species.[17] In one poll, only four in ten people thought climate change was harming them right now, and they thought it would harm future generations, plants and animals, and the world's poor more.[18] A theoretical threat exists "out there" on a different spatial and temporal scale, very different from the typical, immediate risks and threats we encounter in everyday life[19] (whether a hailstorm or an empty wallet).

So why, if most people know about climate change, think it's happening, and it's reinforced by continuing extreme climate-related weather, does it still seem so far-off and unreal? You know some of those reasons from previous chapters. Distancing climate change is a psychological defensive strategy to manage what seems like intolerable anxiety. And, we've been trained to think we've done our part by buying (or not buying) things, rather than getting active at social and collective levels. We also hear lots of arguing about it, so we think, *I'll just wait until some magic consensus arises and someone tells me what I'm supposed to do.* Also, current discourses often do not personally *engage* us in a way that we understand what it means for Eairth and our lives right now or in the future. Let's consider two more factors: how silence can spiral and partitioning.

When I slip a reference to climate change (appropriately) into casual conversation, it rarely receives any acknowledgment. My words fall to the floor or into the bushes, and the conversation moves along as though I never uttered them. People don't want to "go there." Hence, climate silence. Interpersonal communication about climate change is critical and it helps us process the emotions and contradictions we face, yet it isn't happening much.

Climate change isn't the only topic to get the silent treatment; racism, sexual abuse, and genocide come to mind. From these, we know that prolonged and widespread silence has serious consequences. A theory called the *spiral of silence* (posited by scholar Elizabeth Noelle-Neumann and tested for over thirty years) says that when a person perceives that the majority in a social group or collective do not hold the same opinion as she does, she remains silent.[20] The person fears being socially isolated if she deviates from the group norm and majority views. This creates a "spiral of silence" over time, as she (and others with her viewpoint) become less vocal and less willing to express their views. This decreases the visibility of those with differing views, and they appear less numerous over time because their members are silent. Silence heightens invisibility—for climate change and other topics.

The mutually reinforcing spiral of silence makes what's *perceived* as the majority view more predominant and powerful over time. In one study, individuals who feared social opposition from a group were unwilling to speak out – even when they knew the group judgment was factually wrong.[21] In another study, believing that few people in a group shared your concern about climate change contributed to self-silencing.[22] People who have been caught in a spiral of climate silence have told me that they are surprised when they learn climate deniers are by far the minority, not majority, group.

Since we personally experience record temperatures and extreme precipitation events, you would think that climate change would be part of everyone's lived experiences. Yet the discussion of climate change is frequently *partitioned* off from daily life and relevant events, making it seem less real and present in our everyday lives. This happens both in personal conversations and in news stories. Here are some example headlines:

Chinook salmon season cancelled
Hurricane causes gas prices to skyrocket
California deserts, mountains recorded hottest temperatures on record
Hotspots of ocean acidification and warmed waters found

Not one of these news stories connected the dots between the event, related human actions (causes), and their relationship to climate change (effect). None of these stories mentioned climate change, even though there was significant, direct relevance. The salmon season was cancelled because of record warm water temperatures, and ocean acidification is due to oceans absorbing record amounts of carbon dioxide. Even after several years of hurricanes that bore the fingerprint of climate change, news media fail to mention the connection.[23] After the disastrous Midwest floods in spring 2019, ABC, CBS, and NBC failed to mention climate change in twenty-eight stories about them.[24] Just 3 percent of broadcast TV news segments on the California wildfires in 2019 connected them to climate change.[25]

Another example of partitioning was the outbreak of the zika virus before the 2016 Summer Olympics in Brazil. Only a handful of news reports mentioned that the mosquito who carried the virus was a poster-insect for climate change. A documented effect of climate change is that some species (including mosquitos) move north and/or up in elevation in response to warming temperatures. This was a missed "teachable moment" of how climate changes are disrupting individual lives right now.

Also partitioned from climate change were the Central American migrants heading to the US border. Media stories failed to mention that some migration is climate-driven; farmers in Guatemala were starving because a severe drought (in what is normally a predictably rainy country) killed their crops.[26]

For many news stories, climate change is the important "so what" of the story: why is this outbreak or this crisis happening now, and what does it mean for us and the living world? When news stories—and weather reports, and conversations in the workplace, governing bodies, and social groups—partition off climate change, it divorces climate change from daily life. In your own life, turn the partition into a recognition of carbon emissions and climate consequences in all areas of your life, and how the planet's health and your health are linked.

Thus, a key way for conversations to break the silence is to make the climate crisis a *social reality* for everyone you talk to (a concept introduced earlier). Conversations can tell the story of climate change in a way that sounds like it was written for our conversation partners, in language that connects with who they are, what interests them, and what they care about.

AIMS OF CLIMATE CONVERSATIONS

A meaningful conversation about climate change is more than talking "at" someone and telling him what climate change is and why it's important. A conversation is a two-way exchange of ideas, sentiments, observations, and information. It's a fairly informal and usually private talk between two or more people. In good conversations, people listen as much as they speak.

Meaningful conversation is cooperative, that is, you're interested in and open to the perspective of the other, which strengthens and builds relationship. You know you had a good conversation when you felt heard and understood, and that you understood your fellow conversant better. Good conversation is a true exchange of ideas—not just waiting for your turn to talk. If a conversation is more competitive than cooperative and the purpose is to "win" and convince the other person that your perspective is "correct," that's a debate. (The daughter hiking at the mesa wanted to debate.)

Some people use the word "dialogue" to simply mean taking part in a conversation or discussion. Others (such as argumentation scholars) delineate up to six types of dialogue. Here, I present two forms of dialogue relevant to climate conversations: dialogue as collaboration and dialogue that recognizes the role of differences and power.

Dialogue as Collaboration

This form of dialogue refers to a conversation (or series of conversations) in which people with different beliefs and perspectives seek to develop mutual understanding.[27] In *On Dialogue*, David Bohm views *collaborative dialogue* as a specialized form of communication involving consensus, equality, and mutual trust[28] that emphasizes interpersonal win-win relationships. In this way, dialogues are meant to examine (if not suspend) differing assumptions and opinions (to which people are attached and will defend) and for collectively shared meanings to emerge. For example, if several individuals have divergent views about the role of government in protecting public health, they would need to be willing to deeply engage and understand others' sentiments without judgment, while holding their own convictions with humility. Through this collaboration lens, dialogue emphasizes listening over persuading,[29] and embraces a spirit of inquiry that allows participants both to comprehend their world and also transform it.[30] Outcomes from this type of dialogue include opening to different ideas, softening stereotypes, and better articulating one's own values.[31]

Organizations that facilitate and guide formal group dialogue often host multiple meetings with a facilitator, with an end goal of reaching shared understanding, a deep level of trust, and often an ability to move forward on an action or new relationship between parties—be they neighbors, faith groups, or activists. The focus might be on interpersonal, consensual relationship building, with an emphasis on mutual care and vulnerability.[32] Obviously, there is a great need in our world for conversations that help bridge divides that seem too big to span.

Dialogue That Recognizes Power Differences

However, if significant power differences exist in a group of people engaged in conversation, it is possible that something labeled as "dialogue" could be used to manipulate or co-opt participants. Scholars have documented instances of "dialogue" being used covertly by corporations to gain intelligence and divert attention from very real issues.[33] "Stakeholder dialogue" used to manage conflict[34] can sometimes be a sophisticated form of environmental greenwashing.[35]

In these examples, there is a large difference in power among the participants. Communication scholars Shiv Ganesh and Heather Zoller argue that this power differential calls for another form called *agonistic dialogue*.[36] To them, "agonism" (a term used quite differently in different fields) refers to the plural views in a democracy where social conflict is central and necessary. Such conflicts could occur over indigenous rights, environmental degradation, or wealth. Rather than seeking common ground or consensus, agonistic dialogue seeks openness in other ways—such as, being open to alternative stances and counter-narratives, or open to new forms of deliberation. Because democracy emerges out of disagreement and multiple voices, it must pay attention to shifting relationships of power, identity, justice, and social and material needs. To be clear, agonistic dialogue is not "antagonistic." Agonism demonstrates deep respect and concern for the other, whereas antagonism expresses strong dislike or hatred.

An example of agonistic dialogue could be members of a poor, disadvantaged neighborhood conversing with city officials about the additional burdens they face from a climate-fueled hurricane, including unequal treatment by the insurance industry and a lack of city services. City officials hold significantly more power (and purse strings) than the poor residents. Residents wants to make their situation visible and the injustices known to the governing body. Another example might be Native Americans meeting with representatives of an oil and gas company whose activities harm native land; at stake are issues of indigenous identity, social justice, and material commercial benefit.

An important distinction between dialogue as collaboration and dialogue as agonistic is that agonistic dialogue recognizes and foregrounds power differences rather than trying to suspend or avoid them. In this way, conflict and dialogue operate together amidst relationships that unfold and develop over time.

This fits with an earlier discussion of how new narratives that challenge and disrupt the status quo are needed to transform fossil fuel culture. Alternative narratives can supplement or challenge current dominant narratives of nonaction, human domination of nature, and lack of urgency. A risk in collaborative dialogue is that the "common ground" sought may simply recreate existing practices or power relationships that helped spur the climate crisis. Some scholars hold that seeking consensus-oriented views in dialogue can privilege civility over social justice.[37] For example, a powerful alternative narrative to engage legislators in agonistic dialogue might argue that significant action on climate change will not occur until elections are publicly financed to remove undue industry influence.

You might be thinking that agonistic dialogue sounds a bit like activism, and "activist" may not be a label you use to describe yourself. However, remember that an activist is a person who supports strong actions to fix a problem, like climate change. "Strong actions" span an enormous spectrum—from marching in a public protest to talking to your workplace about the climate crisis. Of course, it's up to you what kind of conversations or dialogue you engage in and feel most comfortable with. Obviously, the aims of conversation depend on the partner or group. A conversation with a climate-denying boss calls for a different approach than one with a close friend. The most important thing is that conversations *happen*.

PREPARING FOR CLIMATE CONVERSATIONS

Here are five important ways to prepare for a climate conversation: develop a short climate change definition, make climate change a social reality for your partner, gather local examples, conduct an emotional inventory, and finally create your personal story.

The first step is to develop a short, scientific definition of climate change. The definition only needs to be a few sentences that you have at-the-ready if someone asks you to explain what climate change is. Make it conversational, such as saying "human-caused" rather than "anthropogenic." Make sure the definition includes who causes climate change (humans burning excessive fossil fuels), what happens to the climate cycle when the emissions get trapped (in addition to warming), and how and where we are experiencing climate changes right here, right now. It's also good to remember the high percentage of scientific consensus (97–98 percent) and how very certain scientists are that it's occurring. Practice it out loud. But as you'll see below, some (maybe most) of your conversations don't need to discuss the science at all.

Second, learn how to make climate change a *social reality* for your conversation partner. Since one of the biggest myths about climate change is that it doesn't affect me personally (only distant places or future things), use your conversation to connect the climate crisis with things someone cares about. Perhaps a friend of yours has a son who plays soccer; sadly, the social reality of playing sports outside in increasing summer heat (and air pollution) is very serious. Soccer camps are ending practice early because of extreme heat, and an NFL player died of heatstroke in 2019 after practicing in hot, humid weather. Children overheat faster than adults. And in recent years, much of the world has experienced strings of hottest-months-ever. Use the social reality of soccer, kids, and heat for the basis of a heartfelt discussion with your friend.

If you plan to have conversations with specific individuals you know—a co-worker, neighbor, roommate, or family member—imagine the social reality of each person. If someone is a passionate gardener, runs marathons, is an insurance agent, a foodie, or very active in the community, those interests are relevant to climate change in their lives right now. Health topics are particularly salient, for climate change is the biggest, long-term public health threat of our time. You could talk about risks for pregnant women, heat risks for the elderly, asthma in young children, and mental health issues.

If you begin a climate conversation with someone you don't know, ask questions about what the person enjoys and holds dear to help you understand his/her social reality. Is the person passionate about national parks, nephews, or micro-brews? Even two seemingly different people (politically, religiously, environmentally) share underlying interests (whether a desire for security or the health of family and the planet) that make for a good conversation.

Third, research examples of nearby climate-related impacts. This adds weight to the current social reality in your area. (The National Climate Assessment has chapters by region[38] but try to make it even more localized.) Perhaps heat waves have exacerbated local emergency room visits or deaths. Heat and precipitation shortages may threaten fresh water supplies, such as recently in Capetown, South Africa. If you live in a cold climate, what has happened to winters, winter recreation, and maple syrup? If agriculture is a major regional industry (soybeans, barley for beer, apples, corn), have farmers faced flooding, droughts, increased insects, and so on? Have "green jobs" come to the area – solar, wind power, or electric car manufacturing? What are the positions of local leaders and politicians regarding climate change? One fact I've brought into conversations: Utah is the fifth fastest-warming state, warmed 3 degrees Fahrenheit since 1970, well about the global average.[39] From the same source, I mention to Minnesota friends that Minneapolis and Duluth are among the fastest-warming cities, warmed about 3.7 degrees Fahrenheit since 1970.

Fourth, review table 4.1, *Working through Emotions about Climate Change*. Understand your own emotions and what triggers your denial, apathy, "othering," and so on. It's crucial to remember what you bring to the conversation and to be able to recognize the emotional defenses of others.

Finally, compose your own *personal story* about climate change. How did you come to care and be concerned about it? How has it affected something you care about a great deal—whether outdoor pursuits, the health of someone special, or the future of your kids (or your decision whether to have children). This is the story from your heart about why this issue matters to you; we relate more to a person's stories than solely facts and numbers. Putting your personal story to paper will help you articulate your thoughts and make them memorable to bring into conversations.

LISTENING AND BUILDING TRUST IN A CLIMATE CONVERSATION

For a moment, imagine yourself in the other chair: someone seeks you out to talk about climate change. What would you like that conversation to be like? You wouldn't want it to begin with a science-dump or become an apocalyptic angst-fest, which would make you tune-out and feel uncomfortable. Those approaches suck the air out of what could have been a nice exchange. Obviously, you want to engage in a conversation with someone who puts you at ease and makes you feel that your thoughts and feelings are just as important and valid as his.

One of the best frameworks I've found for *how* a climate conversation should unfold was presented by the Alliance for Climate Education in a four-minute animated video and boldly titled *The Secret to Talking About Climate Change*.[40] The animation demonstrates that a good conversation is more about *listening* than talking, which puts the other person at ease (and disarms potential reactions). A good conversation develops trust and makes the person feel heard. The video recommends that you first think about all the things you want to say about climate change . . . and then *don't* say them! Instead, follow these steps.

- *First, get permission* from the other person to have a conversation. Make sure it's a good time to talk and that you're somewhere quiet and free from distractions; request that both of you turn off cell phones and silence the TV, radio, or music.
- *Next, ask the other person a question* about climate change that taps her feelings or thoughts. "What are your thoughts about climate change? Are you worried about it?" If the person gives a reluctant short answer, ask a gentle prompt: "Why is that?" Then, you just *listen*. Don't judge. Quell the urge to respond and interrupt; let the other person talk at length. Ask follow-up questions that keep the "spotlight" focused on the other person.
- *Reflect back* what you hear the other person say. "So, it sounds like you . . . I'm hearing that you . . . Is that right?" This is called *active listening*. Continue with this focus until the other person feels valued and safe and trusts you.
- *Then, ask the other person if you can tell her/him about your perspective*. Tell your personal story about climate change: why you care, what worries you, and how you became interested and involved. Keep your story personal—no need to mention politics or science if it's not crucial to your story.
- *Finally, ask the other person what she/he thinks about your story, and listen*. Bring the focus back, once again, to the other. Thank the person for talking and say what you learned from her/him.

This kind of conversation creates a safe environment where a person feels free to genuinely express without fear of judgment or attack. It creates trust, builds relationship, and increases the likelihood of future conversations.

SCALING UP YOUR CLIMATE CONVERSATIONS

Think of your conversations as a chance for you and others to reflect, think aloud, and share stories about how you experience climate change and what it means for you. Talk about how it makes you feel, how you feel conflicted or contradictory, where you get stuck or frustrated. Talk about (and have empathy for) what environmental psychologist Renee Lertzman calls the Three A's: anxiety, ambivalence, and aspiration.[41] The first (and perhaps best) step on the way to climate change engagement is conversation; it's how we learn, change, and grow, not just as individuals but also in community.

In your initial conversations, you don't need to discover solutions or mobilize protests or banish meat from the household. Your goal is to empathetically "respond and react to the basic recognition of the current situation facing everyone on the planet."[42] It's not your job to motivate someone to "fix" a problem that is deeply systemic, nor is it your job to "brightside"[43] and tell someone that everything will be just fine. Your conversation is an invitation to be included in a discussion in a way that engages rather than disengages.

If your conversation partner expresses difficult emotions—sadness, anger, shame, or fear—acknowledge and validate those feelings as normal, okay, and understandable. Talking about "negative" emotions with others is actually more influential than not mentioning any emotions at all; one study found that individuals who heard negative emotions thought of the speaker as rational, strong, and caring.[44]

If you encounter a conversant who appears apathetic or ambivalent toward the increasingly urgent news about the climate crisis, remember that this is not the same as not caring or having different or "wrong" values. The messy reality is that all of us have a measure of mixed feelings or contradictions about this serious subject. Just like Al Gore flying around, we know that most of us drive a lot and consume a lot of stuff—at the very same time that we are worried about climate change. Talk about the contradictions of living in a fossil fuel culture, and then ditch the shame! Climate change is not your personal fault nor your lone responsibility.

Getting contradictions and feelings out in the open helps acknowledge how climate change seems overwhelming, insurmountable, and hard to imagine. It's an important point of connection and empathy to share with others. As we explored in the last chapter, empathy is imagining what it's like to be someone else and appreciating her/his dilemmas.[45] Empathy means accepting the other's point of view with understanding and without blame or judgment—the person simply is where she or he is right now.

At times, it might feel like you're walking a tightrope of Fear versus Hope. Things are indeed bad and fear is present, but it's important not to let fear shut you down. We need Hayhoe's "rational hope": understanding the magnitude of the problem and the urgency, but also having a vision of a positive (changed) future with realistically scaled solutions.[46]

Hope and fear are not an either-or dichotomy; you likely feel some of both. Thus, Lertzman relies on a "both-and" or "yes-and" approach: "Yes, things are very bad, and yes, things are likely to get worse; and yes, many people are working on mind-blowing innovative solutions; and yes, humans have tremendous capacities; and yes, it's also really hard and frustrating; and yes, you yourself as a citizen and an individual have a vital role to play in this unfolding mess. And yes, you may feel pretty bummed out at times. If climate change feels hopeless, that's a natural feeling to have. All the more reason to [join in]. You matter."[47]

One place we get stuck, Lertzman says, is when talking about climate change evokes conflicts over our attachments to everyday life and also our desire to be part of the solution: "We can both love and care for our earth, water, and nature, as well as our cars, vacations, and long, hot showers."[48] In conversation, "yes-and" helps us recognize openly that the climate crisis holds a great deal of uncertainty, paradox, and ambiguity, which can actually help build our capacity for resilience in navigating this difficult crisis.

The desire to be part of the solution and make a meaningful difference is an important one; without it, people will tune out to self-protect. The first and foremost thing is talk about it—frequently, and with various people in your life. This helps move you from the level of lone individual to the social level, and it makes a topic "which cannot be named" a front-and-center topic on people's lips.

When your climate conversations have processed emotions and contradictions, turn your conversations toward actions for a positive future. A good call to action to borrow could be Greta Thunberg's consistent themes: *The planet is warming, we are responsible, and we need to fix it. We already have all the facts and solutions to solve the climate crisis—we just need to act. Economic concerns are irrelevant in the face of collapsing ecosystems. If we do not fix this, future generations will remember us for our failures.*

Imagine what you want the future world to be like (more on this in the last chapter). Just as athletes visualize where the ball will land before they launch it, visualize a piece of what your future world would include. We fear change from the status quo in part because it's hard to envision all the components of a change and how we'd then interact in new ways. Start small. Imagine a reformulated relationship with consumer products, or in response to new innovations (such as Project Drawdown). Conversation helps you think through personal (and environmental) priorities, chart a path through the uncertainties, and unfold life in new ways.

Then, scale it up! Envision entirely new models for transportation and for healthier cities—locally, nationally, globally—and how citizens participate and shape democracy. Discuss visions for how institutions and industries could change in the face of climate change—education, agriculture, fashion, architecture, law, an infinite list. Also bring forth ideas for a drastically reformulated relationship between humans and Eairth.

VALUES IN CLIMATE CONVERSATIONS

Another potential avenue to connect with a conversant is to focus on the values important to her or him—the guiding life principles each of us relies upon to help make sense of the world and our place in it. Your personal values are molded through your communal life and interactions with important social others. These values might include what nature is good for or what is considered good and just in society. Your existing values (along with your worldview) filter information you receive, including about climate change.[49]

A values-based approach is popular among some climate communicators, who call for communication that is not "science-down" but instead "values-up."[50] It makes sense that we are more likely to be moved by values we hold dear than by numbers, although some scholars argue that we are moved more by economic and ideological interests and cultural worldviews[51] or by emotions. Nevertheless, values can provide clues in your conversation as to a person's guiding principles and why he holds the attitudes he does.

Shalom Schwartz (social psychologist and creator of the Theory of Basic Human Values) and his colleagues identified fifty-six "universal values" that cluster into categories along two dimensions.[52] One dimension is an "openness to change" versus a desire to "respect or conserve tradition." The second dimension is "self-transcendence" (values that go beyond self-interest such as forgiveness and equality) versus "self-enhancement" (self-focused values such as power, materialism, and success). Other scholars have developed two slightly different value dimensions: hierarchy versus egalitariansm and individualism versus communitarianism.[53]

UK researchers Adam Corner and Jaime Clarke note that conceptual maps about values tend to converge regarding climate change: people who favor self-transcending values are more likely to be concerned about climate change,[54] as are those with communitarian and egalitarian values. Likewise, self-enhancing values, individualistic worldviews, and extrinsically motivated behavior (such as monetary reward) are more commonly associated with the conservative side of the political spectrum.[55]

However, values are not exclusive to just one portion of the political spectrum. In a study with a diverse group of people, a core set of values emerged about energy that included fairness, protection of nature, respect for individual autonomy, future well-being, efficiency, affordability, and avoidance of waste.[56] In a study across the European Union, "communal values" consistently associated with climate change engagement included benevolence (kindness) and universalism (the rights and welfare of all people).[57] Even value dimensions that seem juxtaposed—"conserving tradition" versus "openness to change"—can spur conversation and help find common ground among a variety of individuals. "Human health and welfare" is a universal value used to communicate and engage across differences, including for climate change.

How might values be useful in your climate conversations? If you learn that your conversant is conservative, has a strong national identity, and is concerned about a safe and secure future, those values connect in many ways with the climate crisis: the US could become a leader in reducing emissions and help make the world safer for future generations. If your conversation partner is interested in stability and protection of nature, those values could also be used in conversation.

A caveat mentioned earlier is worth repeating here. If someone is operating solely out of emotions and defense mechanisms, an appeal to her/his values (even deeply held ones) will not be motivating. Even under the best of circumstances, our actions may not spring from deeply held values,[58] especially if they conflict with personal desires. Emotions can circumvent a successful values-based discussion. However, values can also serve as a side door to bring up emotions: *Mom, I know we both care a great deal about nature. We go camping every summer and we all love snowshoeing. That's why I want to talk to you about something that has been worrying me a lot—climate change.*

WHEN TALKING GETS TOUGH

If you know that someone—your uncle, sister, your gym partner—is solidly in the 10 percent dismissive camp and it seems that neither heaven nor Eairth will move him or her from that spot, it's perfectly acceptable not to talk about climate change. Focus on other conversation partners.

For most climate conversations, especially with family members and good friends, again, think of your conversations as a process, not a one-time event. What someone strongly opposed in one conversation may shift upon reflection. A good conversation provides "food for thought" and conversants need time to fully digest and put new or dissonant information and ideas into a new frame of reference. Begin the next conversation with new questions and insights.

If a conversation starts to play out like a never-ending volley of comments back-and-forth, each person is probably focused on making his/her own points but not truly listening or hearing the other person. Stop, and *get curious*. Ask questions to learn what's beneath a particular statement—past experiences, emotions, identity, beliefs? Getting curious helps head off escalation: if you feel your hackles rise, rather than respond, dig deeper. The better you understand the other person—what is important to him and what he identifies with—the better your conversation will be. Listen, and repeat back what you are hearing (active listening).

If a conversation gets too difficult or heated (for either party), it might be better to suspend it, with the understanding that you want to keep the door open to talk another time. Try to reconstruct what happened and how tensions escalated. Perhaps the other person felt ignorant and, therefore, threatened. Maybe you tumbled down the rabbit hole of battling science-denier claims, such as "it's natural," or sun-spots or axis-tilt caused warming. Since climate denial is itself a powerful defense mechanism, it's rather pointless to engage the other's "facts." You could say, "Well, neither of us are scientists, so how about we approach this from another direction," such as lived experiences and concerns for the future. If you asked, "Do you *believe* in climate change?" your conversation may have derailed into politics. If it did, ask to save the political or policy angle for another conversation, and redirect: "Right now, I really want to hear how the climate crisis makes you feel and how you personally experience it in your life." You could also limit the conversation to what climate change means for someone professionally, whether a builder, a rancher, or a teacher.

You may encounter misinformation about climate change, which confuses discourse and is closely linked to skepticism and denial. Misinformation (information that is false, inaccurate, or misleading) and disinformation (deliberately false information spread to deceive and mislead) create doubt about the urgency and science of climate change.[59] On social media platforms, misinformation spreads easily and quickly, and a network of actors (from bloggers to conservative think tanks and media) funds, creates, disseminates, and amplifies falsehoods. If you tweet, you have likely encountered disinformation from an army of automated Twitter bots. A recent study concluded that up to a quarter of all climate change tweets on an average day are produced by mechanized bots.[60] Bots are a type of software programmed to autonomously tweet, retweet, or "like" on Twitter, used to amplify climate denial and claims of "fake science."

Misinformation in a conversation may have come from a person's Facebook friends or Twitter followers. Social networks link you to others who share traits or interests, but such clusters can become "echo chambers" where misinformation reverberates and members may trust false posts and/or be hesitant to correct them due to social norms.[61] Consider it your duty as a social media user to "out" disinformation and misinformation in the best way you can.

When you hear misinformation in a conversation, respond in a way that doesn't threaten identity and that acknowledges the difficulty of knowing what's accurate. Share appropriate knowledge (such as the difference between weather and climate) or direct someone to verified sites or sources to counter confirmation bias. To learn more about strategies to stop the spread of climate change misinformation online, check out this review article.[62]

Beyond disinformation, even faulty arguments can make information seem inaccurate or sidestep points you're making. A classic "strawman" attack is when your argument is misrepresented and then argued from a different direction. For example, you say we should transition to renewable energy, and your conversation partner responds that, well then you shouldn't turn on your computer or drive a car. She changed your topic from how energy is generated to argue that any use of energy is bad. It's important to keep your argument on track and be able to recognize when your partner sends it off-track. Here's a helpful article to learn more about how to deconstruct climate misinformation and identify reasoning errors.[63]

A conversation also can overheat if people believe their identity is being attacked or threatened. This could be true for a highly patriotic or religious person, or for someone whose livelihood is very tied to fossil fuels – bus driver, stock broker, or roughneck on a drilling rig. A coal miner may live in a city built a century ago around mining, and his identity is tightly bound to *being* a miner. Over time, occupations decline or disappear (data entry, travel agent) while new ones appear. But fossil fuel industry workers in particular feel pressure from climate change narratives and likely from climate change itself. It's convenient to blame climate activists for the decline of coal mining and its way of life. If you encounter this in a conversation, first acknowledge and empathize with feeling under attack—no one is comfortable with that. You could express your support of programs that help fossil fuel workers transition fairly to another industry, such as one in West Virginia training coal miners to be farmers.

According to business and sustainability professor Andrew Hoffman, climate change can feel very threatening to a person's economic interests and ideological identity.[64] Climate change might be framed as diminishing citizens' liberties or as an attack on limited government and free markets. You could reframe this in the opposite way ("I think we need to let markets work freely without fossil fuel subsidies"), but that probably won't change a person's attachment to these common and generalized statements. Instead, once again bring this down to the level of personal and lived experience. Ask how the person's own freedoms have been reduced, or how she has suffered economically. When one feels threatened, it's easier to point the finger of responsibility elsewhere ("othering" for self-protection) and ignore contributing factors (such as poor market conditions for coal).

These flashpoints enter the contentious realm of "identity politics" where individuals' complex experiences and identities are crammed into just two political boxes: red or blue. A democracy operates on the basis of plural voices and perspectives, but the current strong partisan divide has led to a "poverty of public discourse" where a person's beliefs become "a set of pre-packaged opinions from partisan shelves."[65] As Daniel Pritchard (a communicator for a nonprofit that fosters dialogue) sees it, "We don't have too much identity in our political lives. We have too little. . . . Over the last three decades, all of our manifold identities—gender, ethnicity, family, religion, ideology—have been consumed by two over-simplified affinities, the only identities that seem to matter anymore in our civic lives: red and blue." Thus, a challenge in your climate conversations is to tease apart Republican/Democrat and conservative/liberal stereotypes into a more personal and unique set of beliefs produced by a person's experiences and values.

It's difficult nowadays to leave politics entirely at the door when talking about climate change, while at the same time it remains extremely important to navigate (if not bridge) the current cultural schism.[66] In such times, the personal touch of a safe face-to-face conversation is truly a gift away from the "noise" of red-blue rhetoric.

A final idea to help navigate difficult conversations is the use of "ground rules" for conversations. Box 5.1 presents Ten Agreements for Climate Conversations; see if it makes sense to share these guidelines with your

---

Box 5.1 Ten Agreements for Climate Conversations

1. I will help find the right time and place to have a climate conversation.
2. I recognize that climate change is an emotional topic, and I respect others' feelings about it. I accept that we are at different stages in the journey to face, understand, and acknowledge it.
3. I will speak for myself, not for others or for entire groups of people.
4. I will seek to understand my conversation partners and their views, not persuade them.
5. I will speak from my own experiences and stay in the present moment (avoiding grand generalizations and bringing in past events)
6. I will examine my assumptions and stay open to questions about them.
7. I will avoid making negative or critical comments about the climate change beliefs or views of a fellow conversant.
8. We will share the time and take turns speaking and listening. I will not interrupt others.
9. I will remain curious and ask honest questions (and provide honest answers).
10. When I hear something that is hard, I will listen as best I can and hang in there.

(Adapted from Living Room Conversations and Essential Partners)

conversation partner before you begin talking and agree to abide by them. You can also craft your own agreements about topics you want left off the table (perhaps religion or economics). The most basic agreements include speaking for yourself (not for generalized others), and speaking about the present moment (not bringing in past blames or hurts). In my experience, if a conversation remains at the level of your personal experience and current emotions, that is beyond arguing; it's simply the way you feel right now.

PRACTICING CONVERSATIONS

Even with these preparation strategies, you might not feel ready to dive into a climate conversation (the first one is always hardest!). If that's the case, you might want to role-play a practice conversation with another person or in a small group. Box 5.2 has three examples.

In my experience, the two hardest conversation tasks are listening and equalizing participation. Listening means attending to what the other says with your whole body, heart, and mind. Listening involves empathizing with the other, curiosity about the other's thoughts and feelings, and understanding the other's views. Listening is not merely quietly waiting for your turn to talk so you can make your points. A typical dynamic in a group (even a small one) is a lot of cross-talk over others' words, which makes it hard to express your thoughts or to feel fully heard and listened to. Use of a timer (during role-playing exercises) signals that it's one person's turn to talk, uninterrupted, while others listen and actively attend. I recommend a small sand-timer; like a "talking stick," as long as you hold it, it's your turn to speak. It also a visual signal of how much time you have left. (I find sand timers far less distracting than cell phone timers, which somewhat removes the speaker and/or time-keeper from the conversation.)

Sand timers are also useful to counter another common dynamic: unequal participation. Some people take up a great deal of air-time and others contribute little, which is true in a classroom, the workplace, casual settings, and even families. A timer encourages the "talkers" to condense and focus their thoughts, and gives "quiet ones" a chance to more fully articulate and expand their ideas without being interrupted. Everyone has important thoughts and insights and deserves the space and opportunity to contribute them.

RESOURCES FOR CLIMATE CONVERSATIONS

It's no surprise that numerous organizations and nonprofits around the world are working to improve and increase conversations about climate change.

Box 5.2   Three Climate Conversation Exercises

1. Facing and Talking about Climate Change (3–6 people)
   - Use a one-minute sand timer. Speak when you're holding it; don't speak when you're not. Keep speaking until the time runs out, then stop speaking. NO cross-talk! After the first person speaks for a minute, reset timer. Repeat until everyone has spoken, then move to the next "go-round."
     1st Go-Around: Talk about how you feel about climate change and the emotions it evokes.
     2nd Go-Around: Talk about contradictions, guilt, or shame you feel as someone embedded in fossil fuel culture who also cares about climate change.
     3rd Go-Around: Respond to others' comments from the first 2 Go-Arounds.
   - Keep the spotlight on others' remarks; do not add your own. (This tests active listening.)
     4th + Go-Arounds: Discuss positive self-efficacy and actions to take as social actors! Think of examples that move away from voluntary, consumer-oriented behaviors. (If you choose not to use the timer for the 4th round, make sure to equalize participation.)

2. Role-playing Conversation with an Uninformed and Disengaged Person (3–4 people)
   - In successive rounds, each individual will have the chance to practice 3 different roles: 1 turn as Uninformed-and-Disengaged, 1 turn as climate change Acceptor, and 1 turn as objective Evaluator (or 2 turns in a group of 4).
   - "Uninformed-and-Disengaged" will role-play a lack of knowledge of what climate change is and a disinterest in engaging – thinking that climate change is distant, in the future, and not relevant to her/his life.
   - "Acceptors" will practice explaining the science of climate change in a conversational and understandable way, and provide local or regional examples of its consequences. Acceptors also will ask "Uninformed" questions about her/his interests and activities in order to make the crisis relevant and connected to the person's lived "social reality."
   - During your turn(s) as Evaluator, describe what you heard overall. Tell the role-players what they did well and what was effective. Provide constructive critique and suggestions.

3. Role-playing Conversation with a Climate Skeptic (3–4 people)
   - In successive rounds, each individual will have the chance to practice 3 different roles: 1 turn as climate Skeptic, 1 turn as climate change Acceptor, and 1 turn as objective Evaluator (or 2 turns in a group of 4).
   - "Skeptics" will role-play the middle-ground (i.e., doubting the science, confusing science with policy, relying on skeptical information sources, and so on).
   - "Acceptors" tap the skeptic's emotions, lived experience, and values. If necessary, steer away from battling science or policy.
   - During your turn(s) as Evaluator, describe what you heard overall. Tell the role-players what they did well and what was effective. Provide constructive critique and suggestions.

This is crucial; we must make room in both public spaces and private lives to express the fear and anxiety we carry and to bring together the personal energies that will carry us into a reformulated future.

Citizen summits, civic expression groups, conversation cafes, and "living room conversations" bring together varied individuals to talk about a particular topic or concern (including climate change), often aided by trained facilitators. For many, it's a chance to talk personally about topics that are largely encountered only from a distance or at the political level.

In the UK, participatory peer-to-peer dialogues are building a sense of broad "citizenship" around climate change, which recognizes societal responsibilities rather than solely individual rights.[67] The Irish government convened its citizens for a national conversation on climate change (and other topics), and the Scottish government held a series of events called The Big Climate Conversation in 2019. A series of citizen assemblies in the UK (a key demand of Extinction Rebellion) will take the form of regional public dialogues about climate decision-making. And the UK organization Climate Outreach through its Global Narratives Project has facilitated climate conversations in Alberta (Canada), India, and beyond.

These forums for environmental citizenship foster fairness and justice among humans and increase participation in local community organizations and decision-making.[68] Individual behaviors matter most as expressions of climate citizenship, rather than as ends in themselves.[69]

Here is a small sampling of places to find guidance, tips, materials, and talking points. They represent a range of formats (individual lobbying to workshops on personal carbon reductions), foci (city-wide action to individual consumer action), scope (individual, city, country), and type of conversation and purpose. No matter the form, talking is good.

- ecoAmerica—This organization builds institutional leadership, public support, and political will for climate solutions in the US and focuses on five sectors: faith, health, communities, higher education, and business. They offer bimonthly climate talking points and a variety of communication guides. www.ecoamerica.org
- Climate Reality Project—This organization is "empowering everyday people to become activists" and leaders to mobilize and fight climate change. They have a guide about climate conversation for families. https://www.climaterealityproject.org/family
- Citizens Climate Lobby—This international grassroots group trains and supports volunteers to build relationships with their elected representatives in order to influence climate policy. They have talking points guides focused on how to talk to your representatives about carbon pricing and fee dividends. www.citizensclimatelobby.org

- Katharine Hayhoe videos—Climate scientist and communicator Hayhoe produced a series of science-oriented videos (*Global Weirding with Katharine Hayhoe*, KTTZ Texas Tech Public Media & PBS, https://www.pbs.org/show/global-weirding/). Also check out her own YouTube channel and website, www.katharinehayhoe.com.
- Carbon Conversations, UK—Carbon Conversations is a psycho-social project about the practicalities of individual carbon reduction while taking account of the complex emotions and social pressures that make this difficult. Groups have convened for over ten years in the UK and around the world. Materials available for download include facilitator guides and workbooks. www.carbonconversations.co.uk
- Climate Outreach—Its mission is "to engage people with climate change from their perspective—not ours." The organization seeks to bridge the gap between research and practice and widen societal engagement. A wide variety of publications and guides are available for free download. www.climateoutreach.org
- Essential Partners (dialogue resources not specific to climate change)—Formerly known as the Public Conversations Project, Essential Partners' mission is to foster constructive dialogue wherever conflicts are driven by differences of identities, beliefs, and values. They offer free training guides and dialogue resources. www.whatisessential.org
- Living Room Conversations (dialogue resources not specific to climate change)—Their conversational model was developed by dialogue experts to facilitate connection between people despite their differences, and to identify common ground and shared understanding. They provide conversation agreements and tips for your own living room conversations. www.livingroomconversations.org

NOTES

1. Not mentioned was that Gore's Climate Reality Project has trained 20,000 people in 150 countries to be climate leaders.

2. A. Leiserowitz, E. Maibach, S. Rosenthal, J. Kotcher, P. Bergquist, M. Ballew, M. Goldberg, and A. Gustafson, *Climate Change in the American Mind: November 2019* (New Haven, CT: Yale Program on Climate Change Communication, 2019).

3. A. Leiserowitz, E. Maibach, S. Rosenthal, J. Kotcher, P. Bergquist, M. Ballew, M. Goldberg, and A. Gustafson, *Climate Change and the American Mind* (New Haven, CT: Yale Program on Climate Change Communication, April 2019).

4. Katherine Hayhoe, "The Most Important Thing You Can Do to Fight Climate Change: Talk about It," *TED Women*, 2018, https://www.ted.com/talks/katharine_hayhoe_the_most_important_thing_you_can_do_to_fight_climate_change_talk_about_it.

5. W. P. Eveland, and K. E. Cooper, "An Integrated Model of Communication Influence on Beliefs," *Proceedings of the National Academy of Sciences* 110 (2013): 14088–95.

6. M. Ballew, A. Gustafson, P. Bergquist, M. Goldberg, S. Rosenthal, J. Kotcher, E. Maibach, and A. Leiserowitz, "Americans Underestimate How Many Others in the U.S. Think Global Warming Is Happening," *Yale Program on Climate Change Communication*, August 2, 2019.

7. Matthew H. Goldberg, Sander van der Linden, Anthony Leiserowitz, and Edward Maibach, "Perceived Social Consensus Can Reduce Ideological Biases on Climate Change," *Environment and Behavior*, June 3, 2019, doi:10.1177/0013916519853302.

8. Goldberg et al., "Perceived Social Consensus."

9. Matthew Goldberg, Sander van der Linden, Edward Maibach, and Anthony Leiserowitz, "Discussing Global Warming Leads to Greater Acceptance of Climate Science," *PNAS*, June 21, 2019, www.pnas.org/cgi/doi/10.1073/pnas.1906589116.

10. Danielle F. Lawson, Kathryn T. Stevenson, M. Nils Peterson, Sarah J. Carrier, Renee L. Strnad, and Erin Seekamp, "Children Can Foster Climate Change Concern among Their Parents," *Nature Climate Change* 9, no. 6 (June 2019): 458–62; Kathiann Kowalski, "Students Can Sway How Their Parents View Climate Change," *Science News for Students*, May 29, 2019, https://www.sciencenewsforstudents.org/article/students-can-sway-how-their-parents-view-climate-change.

11. Karin Kirk, "Changing Minds on a Changing Climate: Reddit Online Commenters Point to Reasons They Went from Being Climate Contrarians to Having Confidence in Mainstream Climate Science," *Yale Climate Connections*, April 18, 2017, https://www.yaleclimateconnections.org/2017/04/changing-minds-on-a-changing-climate/.

12. Kathryn S. Deeg, Anthony Leiserowitz, Edward Maibach, John Kotcher, and Jennifer Marlon, "Who Is Changing Their Mind about Global Warming and Why?," *Yale Program on Climate Change Communication*, January 9, 2019.

13. Helena Bottemiller Evich, "How a Closed-Door Meeting Shows Farmers Are Waking up on Climate Change," *Politico*, December 9, 2019, https://www.politico.com/news/2019/12/09/farmers-climate-change-074024.

14. "Katharine Hayhoe on Why We Need to Talk about Climate Change," *Minnesota Public Radio News*, April 23, 2019, https://www.mprnews.org/story/2019/04/18/katharine_hayhoe_on_climate_one.

15. Leiserowitz et al., *Climate Change and the American Mind: November 2019*.

16. Robert Kirkman, "A Little Knowledge of Dangerous Things: Human Vulnerability in a Changing Climate," in *Merleau-Ponty and Environmental Philosophy: Dwelling on the Landscapes of Thought*, ed. S. L. Cataldi and W. S. Hamrick (Albany, NY: State University of New York Press, 2007), 19–35.

17. A. Leiserowitz, E. Maibach, and C. Roser-Renouf, *Global Warming's 'Six Americas': An Audience Segmentation* (New Haven, CT and Fairfax, VA: Yale University and Center for Climate Change Communication, George Mason University, 2008).

18. Leiserowitz et al., *Climate Change in the American Mind: November 2019*.

19. J. Wolf, and S. C. Moser, "Individual Understandings, Perceptions, and Engagement with Climate Change: Insights from In-depth Studies Across the World," *WIREs Interdisciplinary Reviews: Climate Change* 2, no. 4 (2011): 547–69.

20. D. A. Scheufele, "Spiral of Silence Theory," in *The SAGE Handbook of Public Opinion Research*, ed. Wolfgang Donsback and Michael W. Traugott (Los Angeles, CA: SAGE, 2008), 175–83.

21. Scheufele, "Spiral of Silence Theory."

22. Nathaniel Geiger, and Janet K. Swim, "Climate of Silence: Pluralistic Ignorance as a Barrier to Climate Change Discussion," *Journal of Environmental Psychology* 47 (2016): 79–90.

23. Jim Naureckas, "Media Fails to Tell Climate Story Behind Hurricane Florence," *TruthDig*, September 24, 2018, https://www.truthdig.com/articles/climate-change-made-florence-a-monster-but-media-failed-to-tell-that-story/.

24. Evlondo Cooper, "ABC, CBS, and NBC Completely Failed to Mention Climate Change in Coverage of Major Midwest Floods," *Media Matters for America*, March 3, 2019, https://www.mediamatters.org/blog/2019/03/29/ABC-CBS-and-NBC-completely-failed-to-mention-climate-change-in-coverage-of-major-Midwest-f/223270.

25. Ted MacDonald, "Just 3% of Broadcast TV News Segments on the California Wildfires Connected Them to Climate Change," *Media Matters for America*, November 8, 2019, https://www.mediamatters.org/broadcast-networks/just-3-broadcast-tv-news-segments-california-wildfires-connected-them-climate.

26. Gena Steffens, "Changing Climate Forces Desperate Guatemalans to Migrate," *National Geographic*, October 23, 2018, https://www.nationalgeographic.com/environment/2018/10/drought-climate-change-force-guatemalans-migrate-to-us/.

27. Maggie Herzig, and Laura Chasin, *Fostering Dialogue Across Divides: A Nuts and Bolts Guide* (Cambridge, MA: Essential Partners, 2006).

28. David Bohm, *On Dialogue* (London: Routledge, 2014).

29. Shiv Ganesh, and Heather M. Zoller, "Dialogue, Activism, and Democratic Social Change," *Communication Theory* 22 (2012): 66–91.

30. Paulo Freire, *Pedagody of the Oppressed*, trans. Myra Bergman Ramos and Donaldo Macedo, Thirtieth Anniversary edition (New York, NY: Continuum International Publishing, 2009).

31. Herzig and Chasin, *Fostering Dialogue Across Divides*.

32. Ganesh and Zoller, "Dialogue, Activism, and Democratic Social Change," 71.

33. Judith Richter, *Holding Corporations Accountable: Corporate Conduct, International Codes and Citizen Action* (London: Zed Books, 2014).

34. Mark Moberg, "Erin Brockovich Doesn't Live Here: Environmental Politics and 'Responsible Care' in Mobile County, Alabama," *Human Organization* 61, no. 4 (2002): 377–89.

35. A. Rowell, "Dialogue: Divide and Rule," in *Battling Big Business: Countering Greenwash, Infiltration, and Other Forms of Corporate Bullying*, ed. Eveline Lubbers (Monroe, ME: Common Courage Press, 2002), 33–44.

36. Ganesh and Zoller, "Dialogue, Activism, and Democratic Social Change."

37. J.-N. Kim, and K. Siramesh, "A Descriptive Model of Activism in Global Public Relations Research and Practice," in *The Global Public Relations Handbook*,

ed. Krishnamurthy Sriramesh and Dejan Vercic (New York, NY: Routledge, 2009), 79–97.

38. See https://nca2018.globalchange.gov/.

39. *Climate Central*, "American Warming: The Fastest-Warming Cities and States in the U.S.," April 17, 2019, https://www.climatecentral.org/news/report-american-warming-us-heats-up-earth-day.

40. https://ourclimateourfuture.org/video/secret-talking-climate-change/.

41. Renee Lertzman, "Tackling Apathy and Denial," *UNA-UK Climate 2020*, September 20, 2017, https://www.climate2020.org.uk/tackling-apathy-denial/.

42. Lertzman, "Tackling Apathy and Denial."

43. Adam Corner and Jamie Clarke, *Talking Climate: From Research to Practice in Public Engagement* (Cham: Springer International Publishing, 2016).

44. Brittany Bloodhart, Janet K. Swim, and Elaine Dicicco, "'Be Worried, Be VERY Worried:' Preferences for and Impacts of Negative Emotional Climate Change Communication," *Frontiers in Communication* 3 (2019), doi:10.3389/fcomm.2018.00063.

45. Rosemary Randall, "Loss and Climate Change: The Cost of Parallel Narratives," *Ecopsychology* 1, no. 3 (2009), https://doi.org/10.1089/eco.2009.0034.

46. "Katharine Hayhoe on Why We Need to Talk about Climate Change."

47. Renee Lertzman, "How Can We Talk About Global Warming?," *Sierra*, July 19, 2017.

48. Renee Lertzman, *Environmental Melancholia: Psychoanalytic Dimensions of Engagement* (New York: Routledge, 2015), 146.

49. Wolf and Moser, "Individual Understandings, Perceptions, and Engagement."

50. Andrew J. Hoffman, *How Culture Shapes the Climate Change Debate* (Stanford, CA: Stanford University Press, 2015); Corner and Clarke, *Talking Climate*.

51. Hoffman, *How Culture Shapes the Climate Change Debate*.

52. S. H. Schwartz, J. Cieciuch, M. Vecchione, et al., "Refining the Theory of Basic Individual Values," *Journal of Personality and Social Psychology* 103, no. 4 (2012): 663–88.

53. Hoffman, *How Culture Shapes the Climate Change Debate*.

54. Adam Corner, Ezra Markowitz, and Nick Pidgeon, "Public Engagement with Climate Change: The Role of Human Values," *Wiley Interdisciplinary Reviews: Climate Change* 5, no. 3 (2014): 411–22.

55. Daniel Kahan, "Why We Are Poles Apart on Climate Change," *Nature* 488, no. 7411 (2012); K. M. Sheldon and C. P. Nichols, "Comparing Democrats and Republicans on Intrinsic Values and Extrinsic," *Journal of Applied Social Psychology* 39, no. 3 (2009): 589–623.

56. Karen Parkhill, Christina C. Demski, Catherine Butler, Alexa Spence, and Nicolas Pidgeon, *Transforming the UK Energy System: Public Values, Attitudes and Acceptability – Synthesis Report* (London: UKERC, 2013), http://orca.cf.ac.uk/49200/1/SYNTHESIS%20FINAL%20%28SP%29.pdf.

57. Elena Blackmore, Bec Sanderson, and Richard Hawkins, *Valuing Equality: How Equality Bodies Can Use Values to Create a More Equal and Accepting Europe* (Brussels: Public Interest Research Centre, 2014).

58. Lertzman, *Environmental Melancholia*.

59. Kathie M. Treen, Hywel T. P. Williams, and Saffron J. O'Neill, "Online Misinformation about Climate Change," *WIREs Climate Change*, 2020, doi:10.1002/wcc.665.

60. Oliver Milman, "Revealed: Quarter of All Tweets about Climate Crisis Produced by Bots," *The Guardian*, February 21, 2020, https://www.theguardian.com/technology/2020/feb/21/climate-tweets-twitter-bots-analysis.

61. Michela Del Vicario, Alessandro Bessi, Fabiana Zollo, Fabio Petroni, and Antonio Scala, "The Spreading of Misinformation Online," *Proceedings of the National Academy of Sciences* 113, no. 3 (2016): 554–59.

62. Treen et al., "Online Misinformation about Climate Change."

63. John Cook, Peter Ellerton, and David Kinkead, "Deconstructing Climate Misinformation to Identify Reasoning Errors," *Environmental Research Letters* 13, no. 2 (February 2018): 024018, doi:10.1088/1748-9326/aaa49f.

64. Hoffman, *How Culture Shapes the Climate Change Debate*.

65. Daniel E. Pritchard, "The Poverty of Partisan Identity," *The Fulcrum*, August 21, 2019.

66. Hoffman, *How Culture Shapes the Climate Change Debate*.

67. Corner and Clarke, *Talking Climate*, 83.

68. Andrew Dobson, "Environmental Citizenship and Pro-Environmental Behaviour," in *Rapid Research and Evidence Review* (London: Sustainable Development Research Review, 2010).

69. Corner and Clarke, *Talking Climate*, 85.

Chapter 6

Justice and Faith

The Moral Imperative of Climate Change

Swedish teen Greta Thunberg has galvanized the world's attention. What began as a weekly, one-teen Skolstrejk för Klimatet (School Strike for Climate) by a slight, then-sixteen-year-old female in pigtails on the steps of the Swedish Parliament quickly spread to millions across the globe, particularly young students and teen girls. Her message is powerful because it contains what so many other climate change messages lack: the moral imperative. Here's her remarks to a UN Climate Action Summit in 2019.

> *This is all wrong. I shouldn't be up here. I should be back in school on the other side of the ocean. Yet you all come to us young people for hope. How dare you. You have stolen my dreams and my childhood with your empty words. Yet I am one of the lucky ones. People are suffering. People are dying. Entire ecosystems are collapsing. We are in the beginning of a mass extinction and all you can talk about is money and fairy tales of eternal economic growth. How dare you.... For more than 30 years, the science has been crystal clear. How dare you continue to look away and come here saying that you are doing enough when the politics and solutions needed are nowhere in sight.... You say you hear us and that you understand the urgency.... You are failing us. But the young people are starting to understand your betrayal. The eyes of all future generations are upon you. And if you choose to fail us, I say, we will never forgive you. We will not let you get away with this. Right here, right now is where we draw the line. The world is waking up, and change is coming whether you like it or not.*[1]

Greta speaks of right and wrong, of moral outrage and failed responsibility by the powerful. Greta tells us we have a moral duty to act immediately. Although it's harsh to hear, her critique is important. A moral imperative is missing from much climate change discourse; no amount of information or science will tell us what we *ought* to do.

The new direction and strategy this chapter seeks to add to climate communication is the moral imperative—the responsibilities that we as moral beings[2] have to address this wicked problem. The chapter presents the moral foundations upon which humans rely and how these can be expanded with moral imagination. "Climate justice" clearly invokes the moral dimension: those who contribute least to climate change are generally the most harmed: people of color, women, the poor, and creatures and nonhuman systems. Finally, the chapter examines moral concerns through the lens of religious beliefs. All major faith traditions teach concern for the common Earth; however, some religions (which tend to merge politics and religion) have interpreted these teachings in ways that deny climate change and abrogate moral responsibility.

FOUNDATIONS AND FRAMES: THE MORAL IMPERATIVE

You learned as a young child that it wasn't "right" to hit Fred, even if he took your toy. The Golden Rule taught you to treat others as you want to be treated. Other right/wrong guidelines you learned were telling the truth, keeping your promise, and not harming others.

You didn't learn these morals by yourself but were schooled in "oughts" and "shoulds" by families, teachers, social groups, religious leaders, and communities. "Morals" concern a person's standards or beliefs about what is generally right or wrong, and acceptable or not. "Ethics" are how we apply and use morals in our lives, including determining when human actions wrongly cause great harm, such as to Eairth's climate.[3]

Moral Foundations and Arguments

All moral systems that humans have developed touch on five fundamental categories (table 6.1).[4] These *moral foundations* are believed to be universally present but differently emphasized or culturally relevant.

Table 6.1 Moral Foundations

Fundamental Categories	Opposite "Vices"
Care	Harm, Lack Compassion, Suffering
Fairness and Reciprocity	Cheating, Injustice, Inequality
Loyalty	Betrayal or Disloyalty
Authority and Respect	Subversion, Disobedience, Poor Leadership
Purity and Sanctity	Destruction, Spiritual Corruption, Contamination

Return to Greta's speech and see how she appealed to moral foundations (or their opposite "vices"). Her opening sentence, "This is all wrong," is an all-encompassing moral rebuke. "How dare you" expresses moral indignation at adults' lack of action. Lack of Loyalty is apparent in, "You have failed us" and "betrayal." Stealing her future speaks to Care and Fairness. Two researchers formally decoded Thunberg's moral foundations in this speech with computer algorithms, text data, and evaluations from individuals.[5] They found she appealed to all five foundations but relied more on Care and Fairness. Care language referred to harming the climate and humans through unlimited growth and ignoring the future. Greta invoked Fairness by the failure to adhere to agreements, find solutions, and attend to consequences.

Survey research has found that the moral foundations of Care and Fairness are more associated with "liberals."[6] "Conservatives" associate strongly with Loyalty, Authority, and Purity, although conservatives divide their appeals more evenly across all five foundations than liberals. That suggests that "liberals" could broaden their environmental arguments, particularly the degradation of Purity and Sanctity. Whatever the appeal, moral language must first be disseminated, which makes Greta's worldwide moral narrative all the more important.

To predict people's positions on issues (like climate change), many researchers tend to rely on measures like political attitudes and orientation, demographics (age, sex, education, and so on), or interest in politics. Now, some research suggests that moral foundations may provide a more complete picture about a person's beliefs and can better predict positions on issues. In one study, 25,000 people were asked about their positions on global warming, abortion, defense spending, and same-sex marriage.[7] The moral foundations of Purity, Loyalty, and Care most strongly predicted their positions on these issues—more strongly than political orientation. Care and Purity equally predicted a person's support for tougher measures against global warming. In a second study where people were asked to judge moral issues (like the death penalty, euthanasia, flag burning, and animal testing), Purity strongly predicted "moral disapproval," but political orientation, religious attendance, or interest in politics did not.[8]

It's helpful learn about the kinds of moral arguments into which moral foundations can be placed. Philosophers delineate three types: consequentialist, deontological, and virtue.[9]

Arguments based on the consequences of acting or failing to act are what ethicists call *consequentialist* because it's the action's consequences or results we use to judge whether it's right or wrong (which enhances or harms what we value). For example, we have an obligation to address climate change for the survival of humankind, for children to come, and on behalf of Eairth and its self-regulating systems, from berries and plankton to watersheds and atmosphere.

Moral arguments based on doing what is "right" are called *deontological* or duty-based. The moral duty might be to safeguard the poor or to protect all creatures—a kind of Golden Rule applied to humans and beyond.

Finally, moral arguments called *virtue ethics* are based on who a moral being is and the personal integrity and moral character she/he has. Values that form moral character include compassion (recognizing suffering), love (of Earth's wonders and rhythms), and integrity (consistency between belief and action). Other values (like freedom, patriotism, and duty) could create moral arguments that support action on climate change:

- *It's patriotic to protect our home planet.*
- *It's my duty to God to safeguard creation.*
- *Americans have an obligation to all future families to leave a livable world.*

Princeton philosophy professor Peter Singer called climate change one of the greatest moral challenges of the twenty-first century,[10] which suggests that communicating moral responsibility must be the steady drumbeat accompanying hard changes.

Expanding the Moral Frame and Imagination

At a point in your development, your personal moral standards became what psychology professor Jonathan Haidt calls "fundamentally intuitive, not rational."[11] *Moral intuitions* underlie our belief systems and to an extent our positions on particular issues, even if we don't consciously make the connection. Some psychologists claim there is mounting evidence that environmental attitudes are firmly rooted in humans' moral intuitions.[12]

Yet, climate change may fail to activate people's moral alarm system or generate strong moral intuitions—even among those who know it's a serious problem. Some suggest that the human moral judgment system is not well equipped to identify climate change as an important moral imperative because it's so complex, large-scale, and unintentionally caused.[13] To help, communication strategies can present climate change facts through a moral lens. For example, consider these climate change consequences:[14]

- *Colorful-billed puffin chicks are starving to death because the major food sources their parents always fed them are moving to colder, deeper waters. The adults compensate by feeding their chicks large butterfish, which the chicks are unable to swallow.*
- *Rising temperatures and booming populations of ticks are causing cold-loving moose to move farther north. Warm winters don't kill the tens of*

thousands of ticks that attach to a moose, weakening the immune system and resulting in death, especially for calves.
- *The Bahamas contribute just 0.01 percent to global greenhouse emissions, yet the country faces rising sea levels and intensified storms, such as Category 5 Hurricane Dorian that destroyed several islands and left 600 missing and presumed drowned.*

It's hard to read those facts without an emotional response. Climate change is more than complex scientific findings; it is a science-based problem with profound moral and ethical contours,[15] particularly due to human contributions and responsibility. It's those contours that philosophers Kathleen Moore and Michael Nelson say are largely missing from public discourse:

> No amount of factual information will tell us what we ought to do. For that, we need moral convictions—ideas about what it is to act rightly in the world, what it is to be good or just, and the determination to do what is right. Facts and moral convictions *together* can help us understand what we ought to do—something neither alone can do.[16]

Here are some ways to bring those facts and moral convictions together.[17] First, interpret facts according to moral foundations, such as Sanctity and Purity, Fairness, and so on. Then, stress burdens the facts create, now and in the future. Also, appeal to "emotional sticks" such as hope, pride, and gratitude, which encourage prosocial behavior. Finally, invoke an injunctive norm about pro-planet and prosocial action we should be taking.

> *It's so unfair that we have harmed this amazing Eairth and created such suffering. I'm grateful that I've seen puffins and moose, and I hope my grandchildren will be able to as well. But when I imagine the extreme burdens we are leaving future generations—not just humans but all of creation—I just think, we are better than this. Sure, we didn't intentionally change the climate, but since we learned that we have, it's unconscionable not to take action.*

Although we understand morality from an intuitive and automatic level, we sometimes fall back on rationalist models to compartmentalize relevant, crucial threats: *That's too bad about the puffin chicks and moose and people in the Bahamas, but thank goodness I'm okay.* Rationalization only works if you perceive yourself as wholly unconnected from what happens to life around you. What's needed in this case is an *expanded moral frame* where human health and health of the planet are one in the same.

Another strategy is to practice imagining moral abstractions (whether prosperity, love, fairness) both in your present and into the future. A *moral imagination* allows you to see yourself in the places, for example, where your children and wild creatures will live, and to create empathy for what lies ahead.[18] Imagine the richness of coastal birds that you want young people to see. Imagine your love of the four seasons, now and in the future.

> *I really love moose and it's painful to hear they are suffering; it's so unfair and cruel. It reminds me how important cold winters are, and how winters aren't very cold here anymore either. It's wrong and it's harming us all. We need to act now and keep winters as cold as possible.*

Express moral arguments to your friends, elected representatives, or in the streets. Take them to church, which has a good platform for moral invocations.

Greta Thunberg's strong personal moral compass guides her communication and behavior.[19] If the moral dimension is missing—if it's not part of a personal compass or social norm and it's not voiced or imagined—moral obligations are less likely to extend to actions. Moral discourse can be a powerful addition to climate change communication.

ENVIRONMENTAL AND CLIMATE JUSTICE

It's not surprising that the moral foundation of "justice" has been powerfully invoked to fight climate change. Over the centuries, governments have codified "fairness" and "justice" for the protection of children and animals. Environmental laws regulate some of what we ought and ought not to do regarding the environment, though more often in the name of human health than ecosystem health. "Justice" is most closely tied to the "Fairness" moral foundation, but it has clear ties to all five moral foundations.

The *environmental justice* movement and the newer *climate justice* movement have influenced each other and are fused in many ways.[20] Both are at once local and international and both demand attention to the relationships between human communities and the environments that sustain them. Climate and environmental justice recognize that those who contribute the least to environmental degradation and climate change (directly and indirectly) are disproportionately harmed: people of color, the poor, women,[21] and multitudes of nonhuman species. Core principles of the movement are social justice, democratic accountability and participation, and ecological sustainability.

Environmental Justice: Where People Live, Work, and Play

In 1991, Principles of Environmental Justice were crafted at the multinational People of Color Environmental Leadership Summit. The principles recognized that the oppression of people, the environment, and all species were linked. The first principle (influenced by indigenous attendees) affirmed the "sacredness of Mother Earth, ecological unity, and the interdependence of all species."[22] A key contribution that emerged from this movement was how it changed perceptions of what and where nature and environment are. The environment is not distant and detached from everyday life; it's where people "live, work, and play."[23]

The most powerful predictor of your health is your ZIP code.[24] Why is that? It has to do with where a neighborhood is located, who lives in it, and what pollution it is exposed to. Early colonial settlement (and slavery) and a long history of racial segregation and disinvestment laid the foundation for inequality:[25] poor people and people of color tend to live on less desirable pieces of land, which are closer to environmental contaminants.

In other words, the exploitation of land and people are highly correlated. People of color and the poor are disproportionately exposed to more air pollution, toxic waste, and contaminating industries (trash incinerators, plastics manufacturing, refineries, and coal plants).[26] They also receive unequal protection from harms through existing laws and enforcement than do whiter, wealthier neighborhoods.[27] The early environmental justice movement emerged to fight this unequal distribution of risks and protection, which it considers a form of discrimination: *environmental racism*.[28] Across the globe, environmental justice is a moral imperative to equally protect people from harm.

A recent study added a new twist to the pollution issue by looking at consumption.[29] It found that air pollution is disproportionately caused by white Americans' consumption of services and goods (and the carbon emissions that result from making them), which is then disproportionately inhaled by black and Hispanic Americans. We tend to think of factories or power plants as the pollution source, but polluters wouldn't exist without consumer demand for their products. If you're contributing less to the problem, is it fair that you suffer more from it?

In addition, climate change is a "threat multiplier" not only for heat, climate-linked weather events, and public health and mental health problems but also for *existing* social and economic inequities. For example, a family living in poverty buys few consumer goods, lives in a small dwelling, and perhaps uses public transportation. A woman in rural Africa might care for her children and elderly relatives and must walk to retrieve water and firewood. Both families contribute far fewer greenhouse gas emissions; both families have fewer resources to respond to climate impacts (flooding, heat, or health impacts) and fewer resources to recover from them. In a clear link to the moral trauma caused by slavery, some scholars refer to climate change action as a "carbon abolition movement."[30]

Environmental justice groups typically operate separately from the mainstream environmental movement, which historically has not been diverse in staffing or membership. People of color express just as much (if not more) concern about the environment (and are most vulnerable to environmental harms). Yet, there is a tendency for the US public to misperceive people of color as the least concerned, which researchers label a "belief paradox."[31]

Justice and Climate Disasters

Even though a hurricane or wildfire seems to affect everyone in its path, climate-linked disasters are a polarizing force according to wealth and race, shaping who lives in the most vulnerable areas, who gets hit (versus evacuates), and who recovers.[32] The number affected is staggering: Oxfam International found that climate-fueled disasters forced 20 million people—or one person every two seconds—from their homes every year for the last decade.[33]

Author and activist Naomi Klein argued that wealth and privilege during California wildfires created "climate apartheid" and "disaster capitalism."[34] Rich Malibu residents protected their homes with private firefighters; some failed to inform their housekeepers and gardeners they had evacuated. The California fire protection department relied on prison inmates to do some of the most dangerous firefighting. Migrant farmworkers in wine country worked in the unhealthy haze of wildfire smoke. California's beleaguered private utility Pacific Gas & Electric sold generators off its "marketplace" website to customers who lost power—thanks to PG&E.

Wildfire risk varies according to race. One study found that the people in the western US with the greatest vulnerability to wildfires were disproportionately people of color.[35] Blacks, Hispanics, and Native Americans were 50 percent more vulnerable to wildfires compared to others. The vulnerability included where they lived and access to a car and/or cell service.

Inequities (particularly wealth) also affect how people are able to recover from climate disasters. Wealth inequalities and poverty rates increased after the Camp Fire in Paradise, California, and after hurricanes Katrina in New Orleans and Sandy in New York.[36] If you lack home insurance, you are less likely to rebuild or have insurance payments for temporary housing, food, or to replace lost items. Evacuation and recovery depend on an employer's policies and if you get time off work. One study found that the more aid an area received from FEMA (Federal Emergency Management Agency), the more wealth inequality grew.[37] If you have no insurance and can't access government aid or programs, you lose a lot of money; if you have insurance, you may even make money after a disaster.

Justice, Climate Migrants, and Climate Gentrification

Climate justice impacts also occur from long-term or chronic factors, such as heat, water, migration, and gentrification. In dozens of US cities, low-income neighborhoods are hotter than wealthier ones.[38] Think of any big city: the wealthier neighborhoods have yards with extensive greenery and more parks with shade, while poorer neighborhoods (with more people of color) tend to have more concrete, fewer green spaces, and less shade. US cities with the most disparity between heat exposure and income include Baltimore, Minneapolis, Las Vegas, Los Angeles, and Portland. Las Vegas is the fastest warming city—up 5.76 degrees Fahrenheit since 1970.[39]

There is also a complex connection between climate-related weather and migration—*climate migrants* or "climigrants"—especially from countries dependent on small-scale agriculture. In the poor (and often violent) "northern triangle" of El Salvador, Honduras, and Guatemala, worsening floods and long-term droughts are pushing more and more migrants north.[40] Many displaced Central Americans first relocate within their home countries (a practice that holds true worldwide). Yet if they face violence, marginal employment, and discrimination, they may continue the journey north. Citizens of island nations flooding from sea-level rise also join the ranks of climigrants.

Another longer-term climate justice impact occurring worldwide is *climate gentrification*. Wealthy residents are able to retreat from floods, storms, heat waves, and wildfires and move to safer areas, which then experience soaring property and rental values, squeezing less wealthy residents out. Heat is driving climate gentrification in Flagstaff as Phoenix area residents flee unbearable temperatures (one-third of days are hotter than 105 degrees Fahrenheit and can reach 116). Almost one-quarter of Flagstaff houses are now second homes. As mayor Coral Evans said, "We don't talk much about what climate change means for social justice. But where are low-income people going to live? How can they afford to stay in this city?"[41]

Miami is one of the world's most at-risk places from sea-level rise. Since 2000, many coastal homes have not appreciated in value, largely due to frequent, sunny-day "nuisance" flooding at high tides. Meanwhile, home prices in Little Haiti, a predominantly black neighborhood further inland, have doubled as former coastal dwellers move in.[42]

GOALS AND TACTICS OF A "JUST TRANSITION"

Eight-year-old Levi Draheim lives on a Florida barrier island and worries that his home will soon be under water from rising sea levels. Jayden Foytlin, age eleven, lives in southern Louisiana; her family is still rebuilding from the 2016 storm that dumped eighteen inches of rain in just forty-eight hours.

Oregonian Kelsey Juliana, age nineteen, said wildfire seasons are intense and particulate matter is off-the-charts dangerous.[43] They are three of the twenty-one young plaintiffs in a groundbreaking lawsuit for climate justice. *Juliana v. United States* was filed in 2015; the plaintiffs are now years older.

The young plaintiffs claim that the federal government is endangering their future and violating their generation's constitutional rights to life, liberty, and property. They argue that the government failed to protect public goods under the public trust doctrine (a part of American jurisprudence since the country's founding) whereby a government holds essential natural resources "in trust" for present and future generations, such as oceans and the atmosphere. The Juliana plaintiffs argue that the government's failure to control carbon emissions imperils those public resources in a profound way and violates their constitutional right to a livable climate. They claim that the US government has known about the harmful impacts of burning fossil fuels for over fifty years but has willfully ignored the impending harm.[44] As one federal judge wrote when she ruled that the case could proceed, "Exercising my reasoned judgment, I have no doubt that the right to a climate system capable of sustaining human life is fundamental to a free and ordered society."[45]

The case survived two Supreme Court challenges. Yet, in early 2020, a federal appeals court dismissed the case, the judge concluding that climate change was an issue for the political branches of government, not the courts. The plaintiffs' lawyer plans to appeal.

Litigation is an increasingly used tactic to seek climate justice and provoke action. Since 1990, 1,023 cases in the US and 305 cases in twenty-seven other countries have been filed primarily against governments and major carbon-emitting companies.[46] Youth, investors and shareholders, cities, and states have filed claims. Human rights, rights of nature (discussed in the next chapter), and science (such as event attribution science) play important roles in climate litigation.

Another primary tactic is activism by a burgeoning climate justice movement of global networks and farflung chapters or member organizations. Key action principles include demanding leadership, slowing emissions and fossil fuel use ("leave it in the ground"), protecting vulnerable communities, and ensuring a "just transition" to carbon-free, sustainable communities.[47] Climate justice seeks to hold companies (and/or governments) accountable for impacts and to address injustices to the health and well-being of communities and ecosystems.

Grassroots groups address these issues in local communities and on the global stage. The NAACP (National Association for the Advancement of Colored People) runs its own Climate Justice Initiative. Earth Guardians train diverse youth around the globe to be effective leaders in climate and social justice. Climate Justice Alliance formed "to create a new center of gravity in the climate movement by uniting frontline communities and organizations." Other networks include the Climate Justice Coalition, Indigenous Environmental Network, International Climate Justice Network, and the European-centered Climate Justice Action.

This expansive network is important not just for action but also to participants. The values that emerge from participants' involvement are relationships, accessibility, intersectionality, and community[48] as group members work to create alternative futures. Being involved in such a group also reduces individual anxiety and paralysis about climate change.

Climate justice groups and climate litigation seek three kinds of justice[49]:

- *Intergenerational justice* involves fairness or justice between generations. The *Juliana* lawsuit and the School Strikes for Climate claim it is unjust to allow climate change to cause undue harm to young and future generations.
- *Procedural justice* concerns how disputes are resolved and resources allocated. Climate justice groups demand participation in climate negotiations and decisions about adaptation and disaster recovery. They want to ensure that those disproportionately harmed (including indigenous peoples) are involved and heard.
- *Restorative justice* calls upon the perpetrators of the environmental "crime" of climate change to "repair" the harm done to victims, communities, and ecosystems. Increasingly, restoration is a goal of "victim rights" advocates and used to require environmental polluters to clean up and restore damaged areas. The *Juliana* lawsuit seeks restorative justice through a national restoration plan that would end federal fossil fuel subsidies, phase out carbon emissions, and stabilize the climate system.

Restorative justice has also influenced the international COP (Conference of the Parties) deliberations regarding which countries should bear the greatest burden for decreasing carbon emissions.[50] A key demand of climate justice activists is payment of a "climate debt" by the most industrialized nations (which contributed the most emissions) to the least industrialized. This approach affirms the right of less industrialized countries (which contributed little) to develop out of poverty before contributing significantly to mitigate climate change. Climate justice advocates (and others) who seek a fossil-fuel-free future want to ensure that this enormous shift is truly a *just transition* that supports highly impacted communities and jobs.

If you work in an oil refinery, or your community's main employer is an energy-intensive industry, you may be fearful about a "carbon-free society" and economy. You know a transition to "green jobs" is not a simple, smooth process; it might feel threatening to your identity, sense of place, history, and livelihood. It may not feel like "justice" at all.

In an energy transition, some changes disproportionately impact less-wealthy people. For example, increasing the price of gas reduces carbon emissions (because people buy less gas, a good thing), but the price-hike hurts the poor the most, as well as those in rural areas who drive more miles. A gas price increase spawned intense protests in France (the grassroots Gilets Jaunes or Yellow Vest movement) and in Quito, Ecuador.

Impacted industries, communities, and individuals need support for just transitions. A community dependent on a collapsed fishery (due to climate-warmed waters) needs help retraining anglers for new jobs and attracting new employers to town. In Colorado's North Fork valley, Delta High School integrates solar training into science curricula to give students skill in solar and electrical trades after two coal mines closed that had sustained the area for 120 years.[51]

Communicating with those who work in a climate-changed industry calls for thoughtfulness. According to a report by UK's Climate Outreach, the imagery and language needs to draw from the identity of the affected workers, such as gratitude for the work done in extractive industries that helped build the economies on which we currently depend.[52] It's important to avoid messaging that blames particular industries or workers and instead values their contributions. Even the word "justice" may not resonate. Climate Outreach found that center-right audiences did not respond well to "climate justice" principles such as an obligation to help the world's poorest. Instead, they responded to language of "fairness," "climate and jobs," and changes that can be led by workers, not activists. These audiences appreciated honesty about the challenges ahead instead of simplistic, utopian statements about a shift to green jobs.

Two movements that can be considered part of a "just transition" are food justice and energy autonomy. The local generation of renewable power appears in various cities' climate adaptation plans and is especially appealing to rural electric coops. Food justice focuses on local food production to increase food security and address economic costs that prevent access to healthy and culturally appropriate foods.[53] Community gardens have sprung up in vacant lots and engaged refugee communities and those facing homelessness. Both the food and local energy movements embody revised relationships with the non-human to address climate vulnerability.

The inequities of race, class, gender, and species that climate change exacerbates and makes more visible are deep, systemic problems that were centuries in the making. Tackling these injustices, though difficult and fraught, is a crucial step in a truly transformative cultural transition to address climate change. Such a transformation would shift society's view of risk, the social contract with its citizens, and the world's long-term security. Achieving environmental and climate justice requires many levels of respectful, meaningful conversation and dialogue that give full voice to the moral imperative to address climate change.

THE MORAL IMPERATIVE WITHIN FAITH TRADITIONS

A common feature of faith traditions (also known as religions or wisdom traditions) is that they give individuals a way of knowing and comprehending the larger world—an operating manual of sorts for how to function as part of it. Some religious teachings are just as relevant for secular ethics about how to live in the world. For example, the Christian teaching "As you sow, so shall you reap" reminds us that every action has a reaction, and what we do in the world ripples through multiple generations. In Judaism, kosher rules are designed to inspire thoughtful consumption, "to see that every act of consumption can be sanctified," which asks us to pay attention not only to the products of life but also to process of life.[54] According to Imam Jamal Rahman, climate change motivates us to practice what is at the heart of every religion. First, to become more developed human beings, which involves transforming the ego and opening up the heart. Second, to be of service to God's creation. And third, to realize that we are all interconnected. As the Qur'an says, "So wherever you turn is the face of Allah."[55]

What follows is a brief overview of the spiritual beliefs of five dominant religions and a sixth of indigenous peoples of the Americas. Each overview suggests a moral framing of climate change based on a religion's beliefs and teachings; some frames come from a downloadable guide published by Climate Outreach in conjunction with GreenFaith.[56] To produce the guide, Climate Outreach conducted research (surveys, interviews, and workshops) to articulate each faith's beliefs about the environment, as well as words and phrases that resonate, and language that works across faith groups—a shared interfaith language. They also relied on climate change statements developed by faith traditions. The overview's brevity paints with a very broad brush, recognizing that within a faith tradition, individual denominations or sects have belief systems that differ from the main tradition.

Six Spiritual Traditions' Concern for Creation

Christianity is the largest religion worldwide with nearly a third of the global population adhering to it. There are several strong Christian narratives regarding God, creation, and humans. First, God's creation is a precious gift, a direct expression of the divine, and a gift to us and a source of abundance. For Christians, actions are expressions of faith that bring them closer to God.

Christian morality calls for protection of the vulnerable and innocent (following Christ's teachings). Overall, Christianity believes that the world can be transformed by faith and individuals can be transformed by repentance and forgiveness.

Climate Outreach's research noted that different Christian denominations responded to different narratives about their faith and creation. For some Christians, protection of the vulnerable and innocent was associated with social justice and climate change. New Testament scripture refers to the justice within human actions as a reflection of God: what has been done "unto the least of these my brethren, ye have done it to me" (Matthew 25:40). Evangelicals in particular responded well to the phrase "we are in awe of God's creation," while Anglicans preferred to talk about the "power of love." Likewise, evangelicals desired a greater focus on scripture, while Catholics were more critical of judgmental language.

Islam is the world's second largest religion and includes 23 percent of the global population. Muslims view the world as Allah's creation, and they serve the Creator by caring for creation, which is a gift or blessing.[57] The Holy Qur'an speaks of human beings as appointed by God as "viceroys" or guardians of the Earth, of the heavens and Earth as extensions of God's throne, and of nature as a living scripture and a holy, sacred manuscript.[58] The Qur'an is also critical of waste, declaring those who are wasteful are "brothers of the devil" (Qur'an 17:27). The Prophet Mohammad admonished against wasting water "even if you are by the side of a flowing river" (Sunan Ibn Majah 2:425). Muslims also believe that all actions serve Allah, and all actions have consequences which are answerable to Allah.

In Climate Outreach's research, Muslims responded to narratives about intergenerational responsibility, a responsibility to care for the poor and vulnerable, and the moral challenge of climate change.[59] The natural balance of earth's functions (strongly associated with behavioral rules and justice) is being disrupted by climate change, which can be discussed as "seasonal confusion." Participants also stressed the limits of human power—"all power is with Allah"—although humans can solve the problem of climate change if "God is willing." Muslims see themselves as *khaifah*, or caretakers of creation.

Hinduism similarly has a key narrative of interconnectedness. Hindus believe humans are part of a highly organized cosmic order (*Rta*), and humanity is now living out of balance with our shared planet. For Hindus, the natural world is precious and an expression of the divine; God is everywhere in it—the land, the water, the air. Because Hindus are so interconnected to earth, a Hindu climate declaration says, "We cannot destroy nature without destroying ourselves. Man is integrally linked to the whole of Creation."[60] *Dharma* is an ethical code of duties and eternal principles upon which Hindu morality is based, and which upholds the cosmic order of interconnectedness. A Hindu shows observance of this code through her/his acts. In addition, Hindus believe that an internal change in values is as important as an outward change in behavior. Hindu morality accepts that human acts have wider consequences and can harm others, creating pain, suffering, and violence.

Successful narratives and language for Hindus stress a precious and connected earth, see climate change as a symptom of humanity living out of balance, and believe faith is lived through actions. Because Hinduism has a strong emphasis on spiritual teachers (sometimes called gurus), these teachers and other family role models are trusted individuals to deliver climate messages.

Buddhism had the greatest variation of all the moral faiths, according to Climate Outreach research.[61] For Buddhists, there is no concept of a Creator God or stewardship of creation. Like Hindus, Buddhists focus on interconnectedness, but for Buddhists this is founded on the concept of interdependent co-arising of all things in the universe (with no beginning and no end). Buddhism is a way of life practiced through being and doing, not following a prescriptive faith or scripture. Instead, Buddhist beliefs revolve around the moral principle of respect for all living things. Taking action (and daily practice) is expressed in changing internal values and minds.

For Buddhists, climate change represents an unbalanced world because humans have failed to understand our interconnectedness with the earth. Climate change presents a narrative of moral challenge and the need to extend justice, rights, and compassion to all living things. Because of the centrality of awareness, language can be tied to awakening and taking action instead of silence and lack of awareness.

Judaism is grounded in the cycles of the earth, and most major holidays are associated with harvest festivals.[62] Within rabbinic teachings, Jews are instructed not to spoil or destroy the world, for all life comes from the same place and is connected. Jews believe they are caretakers of Creation, must defend it, and are not allowed to waste or damage it. One's relationship with God must be translated into making real change in the world through moral deeds and through the political and judicial realms. In addition, Jews have pride in their resiliency and believe that through their actions they can restore balance in the harmonious order that was created by God.

Within the shared cultural values of Judaism is a strong sense of a global identity and a commitment to social justice. Thus, climate change is a moral issue of justice and intergenerational rights. Judaism has a strong understanding of the immorality of inaction that is relevant to climate change.[63] The language of renewal is powerful for Jews, for God renews his creation every day. Trusted communicators for climate messages would be rabbis or other respected authorities.

The spiritual beliefs of Indigenous Peoples of the Americas (called Native Americans in the US and First Peoples in Canada) vary by tribe (such as origin stories), but there is an "essential sameness" among them: belief in a living planet.[64] Plants, animals, rocks, air, water, and land are alive, and humans hold an equal and reciprocal relationship with all life-forms. A universal symbol for this is the circle of life or sacred hoop, a shape with no beginning and no end, which represents interdependence and connectedness. Because Earth is a living, conscious being, humans have moral and ethical restraints on their actions. Of great concern are generations yet unborn, thus impact from an action today on the "seventh generation" guides how to live in the present moment. There is rarely a distinction between spiritual and secular life; all is spiritual and sacred, and every action is an opportunity for reflection and contemplation.[65]

Climate Outreach did not research indigenous peoples' beliefs, but the mission of the Indigenous Environmental Network is to protect the sacredness of Earth Mother from contamination and exploitation.[66] Their website notes climate change as one of many issues that affects the survival of future generations; addressing it depends upon tribes' relationship with the natural world and their "responsibility to the sacred principles given to us by the creator."

The Pivotal Role of Religion:
Inner Changes and Outward Actions

As this overview shows, every tradition has principles for human relationships with the earth and all creation, as well as a relationship with a supreme power (or prophet or teacher) and fellow humans. All religions provide guidance and fairly explicit responsibilities for care and stewardship of the living world. The fact that many faith traditions are centuries old also gives them an important "long view" that is significant for climate change.[67] Historical roots link members to the past, ground them in a common ancient history, and provide motivation and inspiration to carry on.[68] Ancient texts fully acknowledge intergenerational and long-term responsibilities, and are capable of imagining nonlinear change and catastrophe. When your spiritual perspective of ethics is so long-term, it contains rich contemplative, devotional, and ritual commitments that help shape enduring values within the religious community.[69] Accepting humanity's limits and limitations is intimately related to the idea of the sacred.

It's no wonder that many scholars and religious leaders believe that faith traditions play a pivotal role in promoting and implementing moral solutions to climate change.[70] After all, major cultural changes have been led by religious voices (not political leaders), including the abolition of slavery and progress in civil rights. The moral authority and institutional power of religions may effect changes in attitudes and politics[71] because their power stems from several sources: moral teachings, religions' functions in society, their considerable reach and influence, and their ability to inspire members to action.[72] Faiths reach 85 percent of the world's population, own a significant amount of land and property, and have huge financial investments.[73]

Those who consider themselves "non-religious" or don't identify with a dominant religion (about one in five in the US) nevertheless may lead what they consider a strong spiritual life. They also may feel strongly guided by personal morals and ethical relationships with others, the earth, and some type of "higher power."

Whatever its source, a spiritual life can help with the isolation, anxiety, and paralysis that many individuals feel about climate change. One's faith can provide a "kind of energetic source and support for sustained outward action that mere ethics and rules can't."[74] Being part of a faith community can allow a person to do "inner work" and find spiritual grounding—and then receive support from the faith community to take outward action.

CLIMATE ACTION BY FAITHS, BUT ALSO TENSIONS

First, let's take a close look at what has been called the most powerful Christian articulation of the moral ramifications of climate change.

Pope Francis' *Encyclical on Climate Change and Inequality*

"Laudito Si," Pope Francis' *Encyclical on Climate Change and Inequality*,[75] has been called one of the most important documents on climate change and a "call to action" along the lines of *Silent Spring* (which exposed the dangers of widespread chemicals and helped launch the environmental movement).[76] In an introduction to the *Encyclical*, scientist and author Naomi Oreskes said two lines of thought particularly stand out:

> The first is an affirmation of our interconnectedness and mutual responsibility toward one another, as well as toward our common Earthly home. The second is a denunciation of the aspects of modern life that have led to our current predicament. The essence of the critique is that our situation is not an accident—it is the consequence of the way we think and act: we deny the moral dimensions of our decisions and conflate progress with activity. We cannot continue to think and act this way—to disregard both nature and justice—and expect to flourish. It is not only not moral, it is not even rational.[77]

Pope Francis says that nature is "a magnificent book in which God speaks to us and grants us a glimpse of his infinite beauty and goodness." He rejects choosing whether people or the environment is more important: "Because human dignity finds its root in our common Creation, caring for our fellow citizen and caring for our environment *are the same thing*."[78]

He states that the biblical instruction to "have dominion over nature" is more correctly interpreted as responsible stewardship: "The biblical texts . . . tell us to 'till and keep' the garden of the world (cf. Gen 2:15). 'Tilling' refers to cultivating, ploughing, or working, while 'keeping' means caring, protecting, overseeing, and preserving. This implies a relationship of mutual responsibility between human beings and nature."[79] The Pope insists that the Scripture offers no justification for misuse or destruction of nature, and instead stresses interconnection: human life is grounded in three fundamental, closely intertwined relationships: with God, with our neighbor, and with the earth itself, which includes all living creatures who possess intrinsic value.

Pope Francis sees inseparable bonds between concern for nature, justice for the poor, commitment to society, and interior peace. He writes of social and intergenerational justice, inequitable harms to the poor, and spiritual impoverishment of the rich. He recognizes that the world cannot address ecological issues without addressing social inequities. He calls for more care and concern for "our common home," and a move from heightened individualism and consumerism to more community connection and dialogue at all levels.

While the *Encyclical*'s release greatly boosted news coverage and conversation, it was not universally embraced in Catholic churches. Pope Francis upset some conservative clerics, though others used the *Encyclical* as a powerful call to action. Undeterred, Pope Francis in late 2019 said he was considering adding "ecological sin against the common home" to the catechism, the book that summarizes Catholic beliefs.[80]

The Growth of Religious Environmentalism and Its Opposition

From Pope Francis' example, and examples from the six faiths discussed earlier, the moral "fit" of religious teachings and climate change action seems very strong. A review article concluded that all religions were promoting some degree of action to address climate change—as part of a God-given duty of earth stewardship—either through individual local actions or in broader (even international) ways.[81] The Episcopal church called climate change denial immoral and said climate action was a moral imperative akin to the civil rights movement.[82] United Methodist, Unitarian, United Church of Christ, and other denominations divested their investments from fossil fuels. Unitarian Universalist and Jewish congregations are involved in state legislative groups and have testified in Senate hearings about climate change.[83] The Church of England recently moved up their zero-emissions target from 2045 to 2030.[84]

In the last several decades, the number and activity of faith-based groups involved with climate change has increased. A Web of Creation directory of "environmental ministry" groups lists 541 in the US.[85] The Evangelical Environmental Network produced a media campaign called "What would Jesus drive?" grounded in New Testament teachings of personal ethics and "love your neighbor as yourself." The Coalition on the Environment and Jewish Life promoted Kosher Kars and Greening Synagogues. Work by Interfaith Power and Light has reached 23,000 religious groups in forty states and across four major faith traditions. Green the Church works with 10,000 black churches to "amplify green theology" and protect God's Creation.

However, not all religions, denominations, or congregations have "greened" or embraced climate change. One study of Christians' environmental beliefs over the past twenty years found a decline in concern (including about climate change).[86] Several factors position faith groups on disparate sides of the climate crisis (particularly in the US): greatly differing interpretations and use of scripture, a close alignment between some faith groups and conservative politics, and a perception of climate change as a liberal environmental issue and, therefore, not a religious issue.

Professor of sociology and religion Laurel Kearns wrote that anti-environmentalists associated with conservative Christianity have been vocal opponents of climate change. She says conservative Christian groups have enormous influence on US public policy (including the refusal to sign the Kyoto climate treaty) because there is "well-funded, well-organized religious opposition to any governmental action, and indeed, to the acceptance of the science of climate change."[87] Kearns identified numerous factors of the conservative Christian hesitancy to respond to environmental concerns:

- the argument that the central focus of Christianity should be on salvation and saving souls, not saving creation;
- a related focus on the individual;
- the charge that religious environmentalists worship creation and not the Creator;
- a hostility or distrust of any science because of creationism;
- an apocalyptic focus on eschatology, or End Times, which seems to predict environmental degradation or collapse;
- a perception that environmentalism is hostile to capitalism;
- and the related accusation that environmentalists are socialists or communists bent on undermining the global economic system.

Some of these positions were evident when a Fox News talk show segment about climate change (spurred by Thunberg's 2019 visit to New York) delivered harsh criticism. The show's hosts said that liberals have replaced religion with climate change, have "forgotten about God," and are "worshiping the environment instead."[88]

Kearns reports that as "religious environmentalism" grows (hundreds of religious nonprofit groups participate in international climate change negotiations), efforts to undermine this work have also increased and caused divisions within denominations.

The Environmental Spilt within Evangelicalism

A good example of the religious division over climate change lies within US evangelicals, where "It has become a wedge issue that has divided evangelicals and broken their once close allegiance with the agenda of the Republican Party."[89] Opposing "sides" employ competing frames regarding climate change and economics, scientific uncertainty, and justice and ethics. Although there is a growing group of "creation care" evangelicals, white evangelical Protestants remain the least likely religious group to say that human activity has contributed to climate change.[90]

Candis Callison, a professor who wrote an ethnography on how climate change "comes to matter," puts a finer point on divisions among evangelicals.[91] Surveys by the Pew Forum on Religion and Public Life divide evangelicals into traditionalists, centrists, and modernists, and labels evangelical Protestants (Evangelical Free, Pentecostal, and Baptist) as distinct from mainline Protestants (such as Lutheran and Episcopal).

In 2006, the "creation care" evangelical voice rose to prominence with *Climate Change: An Evangelical Call to Action*, signed by over 100 key evangelicals as well as key evangelical seminaries. The document's four central claims were (1) human-induced climate change is real, (2) the consequences will be significant and will hit the poor the hardest, (3) Christian moral convictions demand our response, and (4) the need to act is urgent. The declaration encouraged Christians to be more ecologically aware caretakers of creation, to resist wastefulness and over-consumption, and to express humility, self-restraint, and frugality. Creation care evangelicals target the traditional conservative heart of the faith and those not associated with environmental or social justice issues. Climate scientist and evangelical Katharine Hayhoe (mentioned in several chapters) has become a prominent spokesperson for creation care evangelicals.

A difficulty creation care has had moving forward is a sense that "the environment" is a liberal and Democratic matter. These evangelicals stress that creation care is not a new value but an old value of "restoring Eden" that is a biblical mandate and a matter of biblical fidelity.

The most well-known evangelical opposition to creation care is the Cornwall Alliance. In response to creation care's success, the Cornwall Declaration was sent to 37,000 religious leaders,[92] arguing that free-market forces and technology can resolve environmental problems, and climate science is faulty. It reiterates "wise-use" movement beliefs that environmentalism threatens private property rights and capitalism. The Cornwall document interprets key passages in Genesis quite differently, such as if God is sovereign and in charge, global warming must be part of God's plan (which fits end-times scenarios). The Cornwall Alliance has been solidly linked to well-funded, religiously based, free-market movement organizations.[93]

The Cornwall Declaration doesn't just claim scientific uncertainty over climate change; it proclaims that climate change isn't even an issue, thus undermining any discussion of the science. It declares that carbon dioxide can't possibly be a pollutant because it's essential to plant growth. Climate change, like evolution, is "just a theory." "Sound science" become iconic phrasing used by deniers to unseat the veracity of scientific research on climate change.[94]

It is frustrating to people who understand the scientific consensus about climate change to witness such manipulation. But as Callison reminds us, "facts" released into the world have a communal life where meaning and relevance are attached to them. In this way, climate science can be perceived as "ideological" regardless of how vigorously its conclusions hew to a prescribed scientific method. To "believe" a scientific truth as worthy of acting upon may require that the truth be defended by an inside trusted member who provides biblical exegesis to support it. In other words, "Those who deliver the message must be trusted in order to 'bless the facts' and to provide the moral, ideological, and epistemological underpinning necessary to make the claim that action is required . . . and worth prioritizing."[95]

There are competing claims by these factions as to who cares more about the poor. The Cornwall Alliance argues that the costs to address climate change will hurt the poor and undermine our whole free-market economic system. Creation care evangelicals shift the economic frame: poor nations and individuals have fewer resources available and will be hit the hardest, which they couple with the Christian ethic "to protect and care for the least of these as though each was Jesus Christ himself" (Matthew 22:34–40).

Relying on seemingly nonbiblical arguments—economics, scientific uncertainty or bias, and political ideology—seems to diminish what is arguably the strongest religious argument: moral justice. A justice frame strongly fits an apolitical Christian position of loving your neighbor and protecting God's creation.

STRATEGIES FOR RELIGIOUS CONVERSATIONS

Obviously, a call to morals and ethics is not a simple solution for religious engagement in climate change; as Kearns notes, you can't "just add religion and stir."[96] There are, however, effective ways to argue for the moral imperative of climate action in religious terms. One suggestion is to prepare faith-based responses to counter religion-based denial. One example by Episcopal bishop Katharine Jefferts Schori focuses on the moral implications of climate denial; she says that those who reject the science of climate change are turning their backs on God's gift of knowledge: "Episcopalians understand the life of the mind is a gift of God and to deny the best of current knowledge is not using the gifts God has given you."[97]

A broad response was written by English professor George Handley in "Letter to a Student," Clara, who asked him why he as a Christian was so concerned about climate change.[98] Handley is employed by Brigham Young University, owned by the Church of Jesus Christ of Latterday Saints (the Mormons), though his answers resonate with various faith traditions.

First, he explained that "faith" is not about blind acceptance but about asking questions, seeking understanding, and growing morally in light of new revelations. He told Clara it's okay to be "skeptical," which means you need to read widely rather than selectively, and understand how data can be cherry-picked and distorted. A skeptic is thoughtful, inquisitive, and honest about what he knows and doesn't know. Being skeptical is very different from being a "denier"; partisan politics and "beliefs" have gotten in the way of strong scientific evidence.

Handley called a statement such as "God wouldn't let climate change happen, so maybe it's a sign of 'end times,'" "a morally perverse logic."[99] He asked, "why would we want to welcome and even encourage the world's destruction? This is like believing that we should engage in more wickedness so as to bring Christ back sooner . . . [I was] taught that Christians should do the right thing no matter the outcome." When people say, "The world is in God's hands," Handley replied, "Well, yes, but didn't he place it in our charge? If you are a believer in the Bible, weren't we asked to 'take good care of it', to be stewards answerable to our Creator for how we treated the elements? . . . God seems to be pleased with an earth that is flourishing, that nurtures all life, not just our own human lives."[100]

Finally, Handley told Clara that part of addressing climate change is clearly outlined in religious teachings: live modestly, consume only what is necessary, and share generously with the poor. He told her, "Your highest moral principle is not fear but love, proactive love for the poor and for the earth that is our gift."[101]

Handley was a trusted "inside" messenger able to "bless the facts"[102] for Clara. If someone's identity and beliefs are strongly based in a religious faith, it's an effective strategy to "nest" climate change within accepted religious terms. As former South Carolina Republican Congressman Bob Inglis (who lost his seat because of his "belief" in climate change) explained, "If you want someone to have a conversion moment, it helps for them to hear it in their own language, from someone they trust."[103]

An experiment conducted at three evangelical colleges in the United States and Canada demonstrated the power of a trusted messenger to change minds about climate change.[104] After listening to a single lecture on the science of climate change from a self-identified evangelical who embedded a Christian perspective (biblical references, stewardship, and concern for global neighbors) in the recorded lecture, students were significantly more sure climate change was occurring and were more concerned about it. The degree of change was greatest in the most conservative individuals. Even one month later, those changes in belief were still present.

Another important strategy is stronger church leadership and leaders willing to bring climate conversations into church. As Callison noted, Christians are waiting for their leaders to give them permission to care about creation and to say that that's okay.[105] When a reporter toured the American south and asked Christians about climate change, the majority said they never heard climate change mentioned in a sermon, even though many felt a strong moral imperative to protect the planet and its inhabitants—feelings grounded in scripture.[106] Grassroots pressure from congregations might convince pastors to bring the subject into church.

Finally, an important (and lofty) goal is to decouple politics from religion to the greatest extent possible. Particularly in conversations, maintain a focus on the moral edict of a faith tradition for justice and stewardship (rather than yielding to economic and political arguments). Congregants could also pressure their churches to cut ties to the climate denial movement and its funding sources. Institutions of all kinds—including religion—are needed to participate in cultural transformation to address climate change.

THE URGENT MORAL IMPERATIVE

What do we want?
Climate justice!
When do we want it?
Now!

This call-and-response reverberated from the lips of 10 million people around the world in 2019 as they took to the streets demanding action on climate change; a great many marchers were not old enough to vote. Calls for climate justice came from protesters in Nigeria, Antarctica, Philippines, Australia, India, Denmark, Ireland, South Africa, and in over 150 countries and thousands of cities. And it was spurred in large part by Greta Thunberg sitting week after week after week by herself outside Parliament with a protest sign.

The moral imperative speaks to us as moral beings, beyond the science. Gus Speth, an environmental advisor to President Jimmy Carter, said he used to think that the top global environmental problems were things like biodiversity loss and climate change: "I was wrong. The top environmental problems are selfishness, greed, and apathy. To deal with these we need a cultural and spiritual transformation."[107] As Imam Jamal Rahman said, to deal with selfishness and greed, "We have to dig deep into what our religions are really saying and live accordingly."[108]

A survey by Yale and George Mason universities found that five out of six Americans believe it's our responsibility to protect the Earth, rather than simply using it for our own benefit. And among those who believe in God (or believe that the concept of God is relevant to our relationship with nature), majorities of all segments of the "Six Americas" (from "alarmed" to "dismissive") view humans as having a stewardship responsibility, rather than as rulers over the nature. This includes those dismissive of climate change.[109]

This suggests that moral framing – indeed the moral imperative – about climate change could resonate strongly with people, including those who are currently unconcerned. All of us could strive to add messages of Care, Fairness and Reciprocity, Loyalty, Authority and Respect, and Purity and Sanctity to our conversations and actions about climate change and clearly convey the urgent moral imperative of the climate crisis.

NOTES

1. https://www.theguardian.com/environment/video/2019/sep/23/greta-thunberg-to-world-leaders-how-dare-you-you-have-stolen-my-dreams-and-my-childhood-video.

2. Kathleen Dean Moore and Michael P. Nelson, *Moral Ground: Ethical Action for a Planet in Peril* (San Antonio, TX: Trinity University Press, 2010).

3. D. Brown, N. Tuana, M. Averil, and P. Bear, "White Paper on the Ethical Dimensions of Climate Change," in *The Collaborative Program on the Ethical Dimensions of Climate Change* (State College, PA: Rock Ethics Institute, Penn State, 2006), 7.

4. S. P. Koleva, J. Graham, R. Iyer, P. H. Ditto, and J. Haidt, "Tracing the Threads: How Five Moral Concerns (Especially Purity) Help Explain Culture War Attitudes," *Journal of Research in Personality* 46, no. 2 (2012): 184–94.

5. Research by Rene Weber, director of the Media Neuroscience Lab at the University of California, Santa Barbara, discussed in: Tasoff, Harrison, "How Moral Language Helped Greta Thunberg Make an Impact," *Futurity*, September 27, 2019, https://www.futurity.org/greta-thunberg-speech-moral-language-2171622/.

6. Koleva et al., "Tracing the Threads."

7. Koleva et al., "Tracing the Threads."

8. Koleva et al., "Tracing the Threads."

9. Moore and Nelson, *Moral Ground*.

10. Justin Rowlatt, "We Can't All Be Greta, but Your Actions Have a Ripple Effect," *BBC News*, September 20, 2019, https://www.bbc.com/news/science-environment-49756280.

11. Jonathan Haidt, *The Righteous Mind: Why Good People Are Divided by Politics and Religion* (New York, NY: Vintage Books, 2012).

12. Koleva et al., "Tracing the Threads."

13. Ezra M. Markowitz and Azim F. Shariff, "Climate Change and Moral Judgement," *Nature Climate Change* 2, no. 4 (2012), doi:10.1038/nclimate1378.

14. *US Department of Interior*, "9 Animals That Are Feeling the Impacts of Climate Change," November 16, 2015, https://www.doi.gov/blog/9-animals-are-feeling-impacts-climate-change.

15. Candis Callison, *How Climate Change Comes to Matter: The Communal Life of Facts* (Durham, NC: Duke University Press, 2015).

16. Moore and Nelson, *Moral Ground*, xvii (emphasis mine).

17. Markowitz and Shariff, "Climate Change and Moral Judgement."

18. Moore and Nelson, *Moral Ground*.

19. Rowlatt, "We Can't All Be Greta."

20. David Schlosberg and Lisette B. Collins, "From Environmental to Climate Justice: Climate Change and the Discourse of Environmental Justice," *Wiley Interdisciplinary Reviews: Climate Change* 5, no. 3 (2014): 359–74, 370, doi:10.1002/wcc.275.

21. Jimmy Carter and Karin D. Ryan, "How Empowering Women Can Solve Climate Change," *Time*, November 26, 2019, https://time.com/5739622/women-girls-climate-action/.

22. *The Principles of Environmental Justice (EJ)* (Washington, DC: First National People of Color Environmental Leadership Summit, October 27, 1991), www.ejnet.org/ej/.

23. Patrick Novotny, *Where We Live, Work, and Play: The Environmental Justice Movement and the Struggle for a New Environmentalism* (Westport, CT: Praeger, 2000).

24. Robert D. Bullard, *Dumping in Dixie: Race, Class, and Environmental Quality*, 3rd edition (Boulder, CO: Westview, 2000).

25. Eve Bratman, "Development's Paradox: Is Washington, D.C. a Third World City?," *Third World Quarterly* 32, no. 9 (2011): 1541–56.

26. Juliana Maantay, "Mapping Environmental Injustices: Pitfalls and Potential of Geographic Information Systems in Assessing Environmental Health and Equity," *Environmental Health Perspectives* 110, no. 2 (April 2002): 161–71, doi:10.1289/ehp.02110s2161.

27. Robert D. Bullard, *Unequal Protection: Environmental Justice and Communities of Color* (San Francisco, CA: Sierra Club Books, 1996).

28. Bunyan Bryant and Paul Mohai, *Race and the Incidence of Environmental Hazards: A Time for Discourse* (London: Routledge, 2019).

29. Christopher W. Tessum, Joshua S. Apte, Andrew L. Goodkind, et al., "Inequity in Consumption of Goods and Services Adds to Racial-Ethnic Disparities in Air Pollution Exposure," *Proceedings of the National Academy of Sciences* 116, no. 13 (March 26, 2019): 6001–6, doi:10.1073/pnas.1818859116.

30. Eric Beinhocker, "Climate Change Is Morally Wrong. It Is Time for a Carbon Abolition Movement," *The Guardian*, September 20, 2019, https://www.theguardian.com/commentisfree/2019/sep/20/climate-change-morally-wrong-carbon-abolition-movement.

31. A. R. Pearson, J. P. Schuldt, R. Romero-Canyas, M. T. Ballew, and D. V. Larson-Konar, "Diverse Segments of the US Public Underestimate the Environmental Concerns of Minority and Low-Income Americans," *Proceedings of the National Academy of Sciences* 115, no. 49 (December 4, 2018): 12429–34, doi:10.1073/pnas.1804698115.

32. Annie Lowrey, "What the Camp Fire Revealed," *The Atlantic*, January 21, 2019, https://www.theatlantic.com/ideas/archive/2019/01/why-natural-disasters-are-worse-poor/580846/; Phil McKenna, "A Shantytown's Warning About Climate Change and Poverty from Hurricane-Ravaged Bahamas," *InsideClimate News*, September 2019, https://insideclimatenews.org/news/11092019/poverty-climate-change-bahamas-hurricane-dorian-risk-recovery-global-warming.

33. "Climate Change Forces 1 Person from Their Home Every 2 Seconds, Report Finds," *ABC News*, December 2, 2019, https://abcnews.go.com/International/climate-change-forces-person-home-seconds-report-finds/story?id=67446906.

34. Naomi Klein, "Forged in Fire: California's Lessons for a Green New Deal," *The Intercept*, November 7, 2019, https://theintercept.com/2019/11/07/california-wildfires-green-new-deal/.

35. Ian P. Davies, Ryan D. Haugo, James C. Robertson, and Phillip S. Levin, "The Unequal Vulnerability of Communities of Color to Wildfire," *PLoS One* 13, no. 11 (November 2, 2018), doi:10.1371/journal.pone.0205825.

36. Klein, "Forged in Fire."

37. Junia Howell and James R. Elliott, "Damages Done: The Longitudinal Impacts of Natural Hazards on Wealth Inequality in the United States," *Social Problems* 66, no. 3 (August 1, 2019): 448–67, doi:10.1093/socpro/spy016.

38. Meg Anderson and Sean McMinn, "As Rising Heat Bakes U.S. Cities, The Poor Often Feel It Most," *NPR.Org*, September 3, 2019, https://www.npr.org/2019/09/03/754044732/as-rising-heat-bakes-u-s-cities-the-poor-often-feel-it-most.

39. *Climate Central*, "American Warming: The Fastest-Warming Cities and States in the U.S.," April 17, 2019, https://www.climatecentral.org/news/report-american-warming-us-heats-up-earth-day.

40. Gus Bova, "How Climate Change Is Driving Central American Migrants to the United States," *The Texas Observer*, December 3, 2018, https://www.texasobserver.org/climate-change-migration-central-america-united-states/.

41. Oliver Milman, "Climate Gentrification: The Rich Can Afford to Move – What about the Poor?," *The Guardian*, September 25, 2018, https://www.theguardian.com/environment/2018/sep/25/climate-gentrification-phoenix-flagstaff-miami-rich-poor.

42. Jesse M. Keenan, Thomas Hill, and Anurag Gumber, "Climate Gentrification: From Theory to Empiricism in Miami-Dade County, Florida," *Environmental Research Letters* 13, no. 5 (April 2018): 054001, doi:10.1088/1748-9326/aabb32.

43. Steve Kroft, "The Climate Change Lawsuit That Could Stop the U.S. Government from Supporting Fossil Fuels," *60 Minutes*, CBS, June 23, 2019, https://www.cbsnews.com/news/juliana-versus-united-states-climate-change-lawsuit-60-minutes-2019-06-23/.

44. Chaitanya Motupalli, "Intergenerational Justice, Environmental Law, and Restorative Justice," *Washington Journal of Environmental Law & Policy* 8, no. 2 (2018): 333–61.

45. Kroft, "The Climate Change Lawsuit."

46. Joana Setzer and Rebecca Byrnes, *Global Trends in Climate Change Litigation: 2019 Snapshot* (London: Grantham Research Institute on Climate Change and the Environment, London School of Economics, July 2019).

47. Schlosberg and Collins, "From Environmental to Climate Justice," 51.

48. Corrie Grosse, "Climate Justice Movement Building: Values and Cultures of Creation in Santa Barbara, California," *Social Sciences* 8, no. 3 (March 2019): 79.

49. Schlosberg and Collins, "From Environmental to Climate Justice."

50. The Conference of the Parties is the supreme decision-making body of the United Nations and sometimes known by the cities in which they were held each year. COP member countries meet annually to negotiate effective implementation of the Convention, or the United Nations Framework Convention on Climate Change, an agreement that was ratified in 1994 at the Rio Earth Summit. The goal of the UNFCCC is to stabilize greenhouse gas emissions to a safe level.

51. Nick Bowlin and *High Country News*, "How Coal Country Becomes Solar Country," *The Atlantic*, December 14, 2019, https://www.hcn.org/issues/52.1/solar-energy-in-rural-colorado-the-kids-of-coal-miners-learn-to-install-solar-panels.

52. Robin Webster and Christopher Shaw, *Broadening Engagement with Just Transition: Opportunities and Challenges* (Oxford, UK: Climate Outreach, 2019).

53. Alison Hope Alkon and Julian Agyeman, *Cultivating Food Justice: Race, Class, and Sustainability* (Cambridge, MA: MIT Press, 2011).

54. Jonathan F. P. Rose, "A Transformational Ecology," in *Moral Ground: Ethical Action for a Planet in Peril*, ed. Kathleen Dean Moore and Michael P. Nelson (San Antonio, TX: Trinity University Press, 2010), 207–10.

55. Scott Gast, "Climate and Creation," *Orion*, June 2017, pp. 14–21.

56. George Marshall, Adam Corner, Olga Roberts, and Jaime Clarke, *Faith and Climate Change: A Guide to Talking with the Five Major Faiths* (Oxford, UK: Climate Outreach, February 2016).

57. Marshall et al., *Faith and Climate Change*, 24.

58. Gast, "Climate and Creation."

59. Marshall et al., *Faith and Climate Change*, 25.

60. Marshall et al., *Faith and Climate Change*, 22.
61. Marshall et al., *Faith and Climate Change*.
62. Gast, "Climate and Creation."
63. Marshall et al., *Faith and Climate Change*, 27.
64. Jerry Mander, *In the Absence of the Sacred: The Future of Technology and the Survival of the Indian Nations* (San Francisco, CA: Sierra Club Books, 1991).
65. Annie L. Booth and Harvey M. Jacobs, "Ties That Bind: Native American Beliefs as a Foundation for Environmental Consciousness," *Environmental Ethics* 12 (1990): 27–43.
66. https://www.ienearth.org/.
67. Paula J. Posas, "Roles of Religion and Ethics in Addressing Climate Change," *Ethics in Science and Environmental Politics* 7 (2007): 31–49.
68. Ronald L. Johnstone, *Religion in Society: A Sociology of Religion* (London: Routledge, 2017).
69. M. E. Tucker and J. A. Grim, eds., *Religion and Ecology: Can the Climate Change?* (Cambridge, MA: American Academy of Arts and Sciences, 2001).
70. Posas, "Roles of Religion and Ethics in Addressing Climate Change"; Gast, "Climate and Creation."
71. Tucker and Grim, *Religion and Ecology*.
72. Posas, "Roles of Religion and Ethics in Addressing Climate Change."
73. M. Palmer, "Religion Still Plays Vital Part in Struggle for Earth's Future," *Global Times*, November 5, 2009, Posas in "Role of Religion and Ethics" puts the global number at 73 percent.
74. Gast, "Climate and Creation," 18.
75. Pope Francis and Naomi Oreskes, *Encyclical on Climate Change and Inequality: On Care for Our Common Home* (Brooklyn, NY: Melville House, 2015).
76. Naomi Oreskes, "Introduction," in *Encyclical on Climate Change and Inequality*, by Pope Francis (Brooklyn, NY: Melville House, 2015), vii–xxiv.
77. Oreskes, "Introduction," viii.
78. Oreskes, "Introduction," 10.
79. Oreskes, "Introduction," xi.
80. Zoya Tierstein, "The Pope Might Make Destroying the Earth a Sin. Will Catholics Listen?," *Grist*, November 22, 2019, https://grist.org/article/the-pope-might-make-destroying-the-earth-a-sin-will-catholics-listen/.
81. Posas, "Roles of Religion and Ethics in Addressing Climate Change," 15.
82. Suzanne Goldenberg, "Climate Denial Is Immoral, Says Head of US Episcopal Church," *The Guardian*, March 24, 2015, https://www.theguardian.com/environment/2015/mar/24/climate-change-denial-immoral-says-head-episcopal-church.
83. Posas, "Roles of Religion and Ethics in Addressing Climate Change," 15.
84. Katherine Dunn, "For Tackling Climate Change, English Bishops Look to God—and Cathedral Heating," *Fortune*, February 13, 2020, https://fortune.com/2020/02/13/climate-change-net-zero-church-of-england/.
85. http://www.webofcreation.org/religious-education/541-faith-based-groups.
86. David M. Konisky, "The Greening of Christianity? A Study of Environmental Attitudes over Time," *Environmental Politics* 27, no. 2 (March 4, 2018): 267–91, doi: 10.1080/09644016.2017.1416903.

87. Laurel Kearns, "The Role of Religions in Activism," in *The Oxford Handbook on Climate Change and Society*, ed. John Dryzek, Richard Norgaard, and David Schlosberg (London: Oxford University Press, 2011), 414–28, p. 418.

88. H. Alan Scott, "Fox News Panel on Climate Change Says Liberals Have 'Forgotten about God' and Are 'Worshipping the Environment Instead'," *Newsweek*, September 19, 2019, https://www.newsweek.com/fox-news-panel-climate-change-says-liberals-have-forgotten-about-god-are-worshipping-1460310.

89. Kearns, "The Role of Religions in Activism," 425.

90. Megan Mayhew Bergman, "What Would Jesus Do? Talking with Evangelicals about Climate Change," *The Guardian*, December 19, 2018, https://www.theguardian.com/environment/2018/dec/19/talking-with-evangelicals-about-climate-change-south.

91. Callison, *How Climate Change Comes to Matter*, 134.

92. Kearns, "The Role of Religions in Activism," 421.

93. Riley E. Dunlap and Aaron M. McCright, "Challenging Climate Change: The Denial Countermovement," in *Climate Change and Society: Sociological Perspectives*, ed. Riley E. Dunlap and Aaron M. McCright (New York, NY: Oxford, 2015), 300–32.

94. Callison, *How Climate Change Comes to Matter*, 145.

95. Callison, *How Climate Change Comes to Matter*, 154.

96. Kearns, "The Role of Religions in Activism," 425.

97. Goldenberg, "Climate Denial Is Immoral."

98. George B. Handley, "Letter to a Student," *Interdisciplinary Studies in Literature and Environment* 21, no. 1 (2014): 22–32.

99. Handley, "Letter to a Student," 26.

100. Handley, "Letter to a Student," 29.

101. Handley, "Letter to a Student," 31.

102. Callison, *How Climate Change Comes to Matter*, 137.

103. Bergman, "What Would Jesus Do?"

104. Doug Hayhoe, Mark A. Bloom, and Brian S. Webb, "Changing Evangelical Minds on Climate Change," *Environmental Research Letters* 14 (2019), doi:10.1088/1748-9326/aaf0ce.

105. Callison, *How Climate Change Comes to Matter*, 130.

106. Bergman, "What Would Jesus Do?"

107. Quoted by Imam Jamal Rahman in Gast, "Climate and Creation," 18.

108. Gast, "Climate and Creation," 18.

109. C. Roser-Renouf, E. Maibach, A. Leiserowitz, G. Feinberg, and S. Rosenthal, *Faith, Morality and the Environment: Portraits of Global Warming's Six Americas* (New Haven, CT: Yale University and George Mason University, 2016).

Chapter 7

A New Relationship with E*air*th

Imagine a huge halibut—5 feet long and 300 pounds—cruising the coastal waters off Alaska. She's one of the great predators in the ocean, but in addition to prey she ingests micro-plastics. Adhering to many of these tiny plastic bits is methyl-mercury, a highly dangerous neurotoxin.[1] As a top-level predator, her level of methyl-mercury has biomagnified up to 100,000 times higher in her flesh than in the surrounding seawater.[2] Because of this journey —humans discard plastics, oceans break plastics into bits to which methyl-mercury adheres, fish ingest that plastic, and you eat a hunk of that flaky fish – plastic trash has become a powerful new avenue for methyl-mercury to enter your brain and nervous system.

When I learned of this journey, I thought about the phrase my mom would say: "what goes around, comes around." It was good advice for how to treat others, and it's also pertinent for how we treat the living world. So why have we treated it that way—and for so long? Albert Einstein believed it was because humans tend to feel separate from the nonhuman and, therefore, feel unaffected by what happens there, which he calls a delusion:

> *A human being is part of the whole, called by us "Universe," a part limited in time and space. He experiences himself, his thoughts and feelings as something separated from the rest—a kind of optical delusion of his consciousness. This delusion is a kind of prison for his consciousness, restricting us to our personal desires and to affection for a few persons nearest to us. Our task must be to free ourselves from this prison by widening our circle of compassion to embrace all living creatures and the whole of nature in its beauty.*[3]

For Einstein, shifting away from beliefs of separation from nature would be like freeing ourselves from a prison, which sounds like quite an inviting choice. Widening a circle of compassion isn't a technological or political step—it's about choosing a healthier relationship with the entire living world.

The long journey taken by the halibut illustrates that a vital next step in this new era and time is updating our beliefs to recognize that Eairth's flourishing and human flourishing are exactly the same thing. That's how interrelated we are. The climate crisis is actually an outcome or symptom of an even larger crisis: an ecological crisis caused by the unhealthy (and unsustainable) relationship of civilization to the living world—a relationship based on deluded beliefs that we are separate from what keeps us alive.

On a record-warm January day, I heard a man say (while eating trail mix with pistachios), "I'm loving it that winter has been so mild." It was a totally innocent remark; he enjoyed the warm weather. But it exemplified how easy it is (for all of us) to frame our being in the world as solely about Us, forgetting that the world is full of *Others*—non-human beings and elements for whom cold temperatures may be very important. For example, pistachio trees have not been getting enough "chilling hours" in their warming California winter, which has messed up pollination and produced fewer nuts. Cold is a crucial commodity to many Others.

Of all the factors that would help us navigate the climate crisis, shifting from beliefs of separation to beliefs of connection is perhaps the most substantial. This chapter's new direction for communicating the climate crisis examines how communication can assist this shift and help us learn to celebrate our utter interdependence with Eairth. Communicating a new vision of our relationship with and moral responsibility for the larger world—Eairth Citizenship, if you will—is a requisite piece for addressing climate change.

A JOURNEY OF RELATIONSHIPS

A poignant symbol of fossil fuel culture, the climate crisis, and our relationship with Eairth is plastic: pieces of it litter the ocean depths, the Arctic, the landscape, our food, and our bodies. If you are a baby-boomer or younger, your world has always been full of plastic.

Because I wanted to see where and how plastic resided in my own life, I took a "plastic tour" of my house. My conclusion: it's extremely difficult for me to step free of plastic right now—just as it is from fossil fuels. My metal water bottle's top is plastic. So is my driver's license. The outer shells of the radio and TV. In the bathroom, containers for shampoo, lens solution, toothpaste, mascara, cleansers. In the kitchen, condiment and spice bottles, non-scratch utensils, measuring cups, storage containers, coffee grinder, and sandwich bags. CDs and their cases, DVDs, pens, binoculars, jump-drives. A rough estimate: I could avoid or swap out right now at best 10–20 percent for non-plastic alternatives. Now that I had established myself as a member of plastic culture, it was important to understand how we got here, why we remain so tied to and dependent on plastics despite its obvious harms, and how we might shift this relationship—not just with plastic but with the larger world.

The significant use of plastic began in 1945 after World War II. Plastic is made from fossil fuels, so it's not surprising the plastic and fossil fuel industries are deeply connected; inexpensive fossil fuels make inexpensive plastic. It takes a lot of petroleum to produce plastic: the oil needed to make a plastic water bottle would fill it a quarter-full.[4] Plastic is a "forever" product in the environment: it breaks down into smaller and smaller pieces but *never* goes away; in sea water viewed through a microscope, next to the plankton you will see hundreds of colorful flecks of plastic. Despite heightened awareness of plastic pollution (in water and on land), plastic production is projected to *quadruple* by 2050.[5] The 165,000,000 tons of plastic in the oceans now is as though one garbage truck dumped a full load of plastic into the ocean every minute.[6] By 2050, plastic in the ocean will *outweigh* the fish.

Until we discard it, plastic is incredibly handy: its forms range from hard and thick (car-parts, zippers, printers) to soft and thin (kitchen wrap, phone screen protector, "disposable" wipes). The carbon footprint of plastics, cradle-to-grave, is enormous. The Center for Environmental Law gathered global data for greenhouse gases produced during extracting, manufacturing, shipping, and disposal (such as incineration) and estimated that plastics could account for 56 billion tons of carbon emissions between now and 2050—fifty times more than the annual emissions of all coal power plants in the US.[7]

Much plastic use is one-and-done use; a whopping 80 percent of our trash consists of things we used just once.[8] A *year's worth* of plastic bottles stacked end-to-end would stretch to the moon and back sixty-five times. Even though I routinely "recycle," I recognize it's not a solution to the plastic littering our world and bodies.[9] In 2018, China stopped buying the 7 million tons of plastic it imported and recycled—in part, because the material was too contaminated with food, drink-caps, paper, and unrecyclable plastic (such as plastic wrap and plastic bags). Perhaps your city is one of those across the world that is now rethinking (or halting) its recycling program. Some cities have taken baby steps by banning plastic grocery bags or planning to phaseout single-use plastics.

Plastic gets into the oceans from direct dumping and littering, and from storm-water runoff and treated wastewater. One study[10] calculated that rain washed 7 trillion pieces of microplastics (much of it particles worn off tires) into San Francisco Bay each year.[11] (Microplastics are very small plastic pieces, from the width of a pencil eraser to less than a human hair.) The fibers from tires amount to 300 times more than the plastic microfibers from laundered polyester clothes and other plastics washing down sinks and sewers. Another study found that 365 microplastic particles per meter fell on the Pyrenees Mountains in southern France; half of the particles were not visible to the human eye and entered the human body through the nose or mouth. The researcher labeled microplastics a new atmospheric pollutant.[12] Even tinier "nano-plastics" can easily cross barriers, including placental and blood-brain.[13]

When a creature ingests plastic, its body responds to it as a foreign stressor. Ingested plastic can block a fish's gills, cause lacerations, and alter behavior.[14] Pieces of plastic (and plastic bags) kill a wide variety of creatures—birds, sea turtles, and marine mammals—by restricting airways or digestion.[15] A study of a local fish market in Half-Moon Bay, California, found that one-quarter of its fish and one-third of its shellfish had plastic particles in the flesh.[16]

We eat microplastics in mussels, fish, chicken, and honey[17] and ingest them from bottled water and from the air. A scientific review of fifty-two studies concluded that humans on average "eat" a credit card's worth of microplastic *each week* (about five grams), or half a pound a year.[18] The European Union wants to classify microplastics as a contaminant that is unsafe at any level of discharge. Plastic often contains harmful chemical additives: plasticizers, dyes, and flame retardants, though few studies have quantified the health impacts of ingesting them. Microwaving food in plastic is a pathway for plasticizers to leak into your food; bisphenol-A (BPA) makes clear, hard plastic and phthalates make plastic soft and flexible. Both are believed to be "endocrine disruptors," substances that "mimic human hormones, and not for the good," according to Harvard Medical School.[19]

And then there's the mercury. Two-thirds of the mercury in the ocean comes from human activities; the largest source is burning fossil fuels (especially at coal-fired power plants). Some mercury comes from garbage incineration, mining, other industrial activities, and even silver dental fillings. When mercury reaches the ocean, microorganisms convert it into much more dangerous methyl-mercury, which naturally adheres to trillions of microplastics.[20] Scientists have found that microplastics increase methyl-mercury concentration in the gills and livers of fish where it accumulates.[21] The effects of methyl-mercury on humans are well-documented: damage to neurological development and the central nervous system, and cognitive impairment, especially in children. About a billion people worldwide rely on fish for protein.[22]

After reading all these studies, I was deeply disturbed. It was hard to imagine that these handy, cheap plastics I used and threw "away" (or wore "away" like tires and tennies) were harming me and Others. I kept returning to "what goes around, comes around." How on Eairth did I imagine that such actions would *not* harm me? Just like the song we learn about human anatomy—"ankle bone's connected to the shin bone"—the connections between our lives and everything that surrounds us should not be surprising; if anything, they are expected. We breathe the planet's atmosphere, eat food grown in soils, and drink from watersheds. It's not enough to quit the bottled water and deli containers. To thoroughly eliminate my weekly consumption of a credit-card's-worth of microplastics requires a substantial shift of underlying human beliefs from ones of separation to ones of connection.

In his 1974 landmark paper, "Resources as a Constraint on Growth," economist William Nordhaus argued that we need to change from a "cowboy economy" (a highly human-centered one where humans use resources profligately) to a "spaceship economy" (where "free goods" like air, water, quiet, and natural beauty receive the same care as scarce or private goods).[23] What he describes as an economic transition also describes a transition of beliefs—from one where humans are the dominant cowboys, to one where humans operate as interdependent and humble components of a finite planet. Let's first look closely at those "cowboy" beliefs.

DANCING AT THE TOP OF THE PYRAMID

"Earth Overshoot Day" marks the day when we've collectively consumed the amount of food, timber, fiber, water, and carbon that the planet can produce or renew in a year. As the name suggests, it's a barometer for how we're stressing the planet's elements, species, and ecosystems, and are out of sync with what Eairth can sustainably provide.

In 1970 (the year of the first Earth Day), Earth Overshoot Day fell on December 29. If we were living "sustainably" it would fall into the following year or beyond; it's important to keep healthy earth elements "in the bank" for catastrophes and for a rapidly growing population. In 1999, Earth Overshoot Day fell on September 29, and twenty years later in 2019, if fell on *July 29*. That means the world's people use 1.7 Earths each year, and it takes the planet twenty months to regenerate what we use in a year. If everyone on the planet lived an American lifestyle, that would require a whopping five Earths a year.[24]

This communicates that humanity is living unsustainability and carrying an ecological debt of enormous proportion. The overshoot occurs because of the current political and industrial ways of living that humans have constructed, which is driven by underlying dominant beliefs toward Eairth. Two scholars laid out some of these deeply rooted beliefs:[25]

- The Earth exists merely to provide for human comfort.
- We are free to consume or destroy natural "resources" at will but are safe from destruction ourselves.
- The extinction or reduction of other species does not matter.
- The Earth will continue to sustain us, even if we do not sustain the Earth.

These are the beliefs of *anthropocentrism*, a deeply human-centered belief system. It views humans as having more value and importance than all other beings, and maintains that humans can control and dominate the non-human world. It's a hierarchical system with humans on top and superior to all Others. In essence, we're dancing at the pinnacle of a giant pyramid with mostly utilitarian regard for non-human Others.

Anthropocentrism is not a new belief system; it can be traced to various points in human history, including the Great Chain of Being (in Latin, *Scala Naturae*, or Ladder of Being), a concept derived from Plato and Aristotle and depicted in a drawing with humans on top above the rest of life (primarily other animals). Some biblical interpretations conclude that humans have "dominion" over the earth (although other interpretations label this "stewardship"). *Human exceptionalism* maintains that humans are somehow better and more morally relevant than all Others, that all human behavior is controlled by culture and free will, and that all problems can be solved by human ingenuity and technology. There isn't room in this chain of being for our behavior being dictated, for example, by insects, winter, or drought. For non-sentient elements—air, water, climate, sun, soil and substrate, and plants—anthropocentrism attaches only instrumental or *utilitarian value*, even though these elements keep us alive.

Historically, this has been a societal blind spot: a failure to recognize the deep interdependence between human-made systems and the living systems that sustain us.[26] (An exception to this are the beliefs of various indigenous populations, which we'll return to shortly.)

A Growth Economy on a Finite Planet

The systems and institutions we have constructed—such as economic, governance, and law—reflect this perceptual gap between humans and the living world. I heard an advertising slogan recently from a global currency and "futures" trading company: *It's Your World—Trade It*. This communicates anthropocentric thinking: the world belongs exclusively to humans and humans should trade it for "futures." For example, our growth-based economy is built on the assumption that precious materials—fresh air, clean water, metals and minerals, and species—"can magically regenerate themselves in an instant, that somehow the Earth will expand to meet our insatiable appetites," said UK Member of Parliament Caroline Lucas.[27]

The premise of a growth-based economy is that each year we will produce and consume more than the previous year, ad infinitum; this assumes that healthy ecosystems will produce not just the same amount every year, but ever *more*. Yet, fewer fish are available for humans (not to mention for

Others) because fishery numbers are declining in most of the world's rivers and oceans. Our economic system (growth-based capitalism) places value in the product (fish) and the transaction (selling and buying fish), but no value in the vast ecosystem that created and supported that fish. As urban ecologist Jonathan Rose warns, "There is a disconnection at all scales between the things that every culture claims to deeply value, and the values we express in our economic structures and daily actions. The consequence of this disconnect is not conceptual; it is fatal. We need to connect what we ecologically know with what we value."[28]

Some scholars attribute such a mechanistic worldview—that nature is the machine and humans the engineers—to the tremendous advancement of scientific knowledge since the Industrial Revolution. "Objective" scientific knowledge prompted humans to believe they could remake and master the world. Political theorist Bob Tostevin claimed that once science uncovered scattered pieces of what makes living systems or organisms tick, it was easy to forget how very independent the living world is from human will, and easy to conflate the control of nature with science.[29]

Anthropocentrism views humans as the *subjects* and the world (and all in it) as an *object*, something humans are able to manipulate and control. For example, consider how "land management" demonstrates this view of the world:

- A lawn management company reduces your yard's problems to "weeds," "pests," or "nutrients"; the lawn is a passive object, not a community of living beings.
- Large mammals, both predators and "game," are controlled through hunting, removal, and "habitat manipulations," even though ecosystem health depends on the health of all plant and animal species within it.

According to author and mathematician David Orwell, "We will choose to protect nature only if we value it—and not just as an object, but because it is alive. The only way we will respect it is if we understand that we cannot control it."[30]

Anthropocentric beliefs are so dominant and widespread that they appear second-nature, a way we move through the world that seems utterly normal. That's part of the problem; "normalcy" reinforces that we are separate from and untouched by harms to creatures and systems upon which we depend for everything. When air quality worsens in Salt Lake City, officials say the air is "unhealthy for sensitive groups." This warning communicates that most everyone (human and non-human) is safe from harm, in spite of voluminous scientific research that even *low* levels of air pollution contribute to heart and lung disease, autism, dementia, and pregnancy and birth outcomes.[31] Beliefs of separation also prop up social inertia over climate change: we know and see the climate changing but hold onto the delusion that it is not really harming *us*.

Climate-related weather disasters challenge the notion that humans control and are separate from the living world. Amitav Ghosh noted that the uncanny and improbable events (of climate change) that are beating at our doors seem to have stirred a sense of recognition, "an awareness that humans were never alone, that we have always been surrounded by beings of all sorts who share elements of that which we had thought to be most distinctively our own."[32] Climate change has given us a renewed recognition that entities like forests or rivers are fully capable of inserting themselves into our lives and thoughts.

Each day, we communicate human-centered beliefs in hundreds of simple actions. It's "cheaper" to replace rather than repair. More pounds of chemicals are applied to lawns than to agricultural fields. A sit-down restaurant where I joined a friend for lunch used only plastic "dishes." Existing trees, plants, and ground cover are usually removed before a building project. Roads of all sizes split habitat, block movement, and kill all manner of creatures—owls, salamanders, badgers, deer. We leave doors and windows open while the furnace or air conditioner run. For the privileged, the size of cars increases, as do house sizes and the number of products inside. When we're done with something, we throw it "away."

You might be thinking, "Yes, that's what we do, but it's just modern life." Yes, it's modern life on a planet now in serious trouble. In an era of climate change, we are doubly aware of the extended relationships each of us has with the nonhuman living world, and that truly, "what goes around, comes around."

For example, consider all the relationships embedded in the food on your dinner plate. Your chicken may have come from big chicken-producing states like Arkansas or Alabama, a region of the country that has experienced warming as well as record flooding (which spreads animal waste). Perhaps Nebraska or Iowa farmers grew the grains that your chicken ate, where recently fields were too flooded to even be planted. Now imagine the potatoes and lettuce on your plate, growing in soil in Wisconsin or Arizona. The plants were visited by insects, and birds who ate the insects, and also bees who pollinated the plants. All pollinators are impacted by a changing climate: 40 percent of invertebrate species are in decline and bird populations have declined by 3 billion birds over the last fifty years.[33] If you ate shrimp last night, you know from previous chapters that oceans are warming and acidifying (faster than any known change in ocean chemistry in the last 50 million years[34]). Although you value these foods in a utilitarian way for how they satiate and please your palate, you have relationships with each foodstuff, itself woven into ecological communities of great value.

Switching Fuels and Switching Beliefs

The current discourses of climate change often portray the solution as switching out fossil fuels for renewables. That would be fantastic, of course! It would limit future warming (although weather-weirdness would continue and even increase) and we could turn more communication and attention to adapting our lives and cities to a climate-changed world.[35]

But consider this: a transition from fossil fuel to renewable energy *by itself* won't address the myriad problems we've been discussing: quadrupling of plastic, methyl-mercury in fish and human bodies, the precipitous decline of birds and insects, and so on. Why not? If our anthropocentric beliefs and practices continue in the same way and at the same pace—just with renewable energy—humans would still consume the Eairth at an untenable pace. If humanity believes that "the earth will continue to sustain us no matter what we do," it could well be that renewables (not fossil fuels) will be the energy that leads to the collapse of oceans, agriculture, and so on. Too often, "sustainability" implies sustaining as much as possible of our current *unsustainable* way of living, only with wind turbines and electric cars. But as Earth Overshoot Day demonstrates, we are already far beyond sustainable.[36]

As noted, climate change is a symptom of a larger ecological crisis, brought about by our relationship of separation from the living world. The COVID-19 pandemic is another example of that relationship. Even though this coronavirus is referred to as a "public health" issue, many researchers say the pandemic underscores the need for a "planetary health" approach: human, animal, and environmental health are interconnected.[37] Zoonotic disease transmission from nonhuman animals has existed for centuries, but scientists say this spillover was because of human actions, not the animals. A principal driver of spillover is land use change and habitat disturbance—principally, biologically diverse rainforests removed by logging, cleared for agriculture, and urbanized. Degraded, fragmented habitats, booming human populations, and a global wildlife "trade" put stress on animals and bring them in closer contact with humans, making virus transmission likely. In this dynamic, climate change is a threat multiplier: shrinking and unhealthy animal populations face extreme weather events, altered temperatures, and are forced to migrate to new areas nearer humans. One scientist said COVID-19 is humanity's first clear, indisputable sign that environmental damage can kill humans fast, too.[38] Some scientists fear that the next emergent pandemic could originate in the Brazilian Amazon where deforestation has reached its highest level in a decade.[39]

Tools used to fight the pandemic have been those of health technology. Modern humans put great stock in technology's ability to improve (if not save) human lives, whether due to disease or climate change. Almost half of Americans think new technologies can solve global warming without individuals having to make big changes in their lives.[40] Some scientists believe that technological interventions can fix even the largest societal problems and can "solve" climate change through "geo-engineering." The trailer for Leonardo DiCaprio's climate change movie *Ice on Fire* ends by saying, "Technology can save the world."[41] This catchy conclusion (and geo-engineering generally) misjudge the nature of this problem: climate change lacks simple fixes because the trillions of interactions and interdependencies in the living world are beyond human comprehension, let alone control.

As discussed previously, climate change is a cultural problem as much as an energy or science problem. Ecosystem degradation (which has occurred for several thousand years but has now severely peaked in intensity) arises from the basic relationship between civilization and nature.[42] As activist Rowan Williams put it, "We have completely and successfully internalized the belief that the world is made up of dead stuff. . . . We have lost or suppressed any memory of what it is to live in alignment with the rest of the world."[43]

This is why climate change is a symptom of a larger ecological crisis that "carbon-free" alone can't solve. What will solve it is shifting from anthropocentric beliefs to seeing ourselves in interrelationship with the living world.

MENDING THE CIRCLE

The myth of separation[44] contributes mightily to the exceptional passivity toward the environmental harms we witness or suffer from each day, whether air pollution or toxics. Transforming beliefs of separation is not about undertaking a particular behavior or embracing a certain product or "lifestyle"; it's a whole new orientation that internalizes and celebrates our unquestionable interconnectedness with all that lives with and around us. Instead of dancing at the top of a pyramid, we dance within a circle of all life.

The circle represents *ecocentrism* (also called *biocentrism*)—a nonhierarchical mix of interdependent relationships or a web of all life.[45] Humans are an integral and interconnected part of the biological world but no more or no less important than other elements. All living and nonliving entities in the world—animals and birds, insects and soil, water and rocks, air and plants—have *intrinsic value*. Intrinsic value means a rock and a beetle have inherent worth just because they exist—regardless of whether they benefit humans or are considered economically valuable. The Society of Conservation Biology affirms the importance of "intrinsic value in the natural diversity of organisms, the complexity of ecological systems, and the resilience created by evolutionary processes."[46]

Ecologist and author Aldo Leopold said humans must move beyond thinking of the living world solely in utilitarian or economic terms. Because all life possesses intrinsic value, humans must fully consider the *holistic integrity* and healthy functioning of entire ecosystems when they take actions. His "land ethic" states: *A thing is right when it tends to preserve the integrity, stability, and beauty of the biotic community. It is wrong when it tends otherwise.*[47] Preserving holistic biological health is the overriding, compelling principle, regardless of present or future benefits people might receive from certain organisms or materials.[48] As one biologist concluded, "That which constitutes a good life for those of one species ought not to compromise the good lives of those of other species with [whom] we share the planet."[49]

Valuing a river system intrinsically is different from attaching economic value to the benefits the river provides humans—sometimes called "ecosystem services."[50] If a river's "services" were quantified as sediment filtration, water potability, or carbon sequestration, think of all the other values (and inhabitants) that might be neglected. Likewise, attaching monetary value to the carbon sequestration performed by a forest ecosystem underrepresents or undervalues the myriad intertwined lives within that ecosystem. It would be like valuing your attendance at work but ignoring your value as a person and all you give to other employees and clients.

The good news about living in this large circle of intertwined life is the amazing enrichment it brings—for me, joy at a vivid sunset or a rising moon, happiness watching otters play, delight in birdsong at dawn, or serenity at the scent of spring carried by a warm wind. Many people maintain that our love of the living world is already deeply embedded within us all. *Biophilia* is the innate tendency to seek connections with nature and to love all that is alive.[51] Biologist E. O. Wilson claims that biophilia has ancient roots as the instinctive bond between humans and living systems, and we subconsciously seek these connections. At an immemorial, core level, humans' natural love for life helps to maintain all life. Human cultures worldwide possess what could be called "a spiritual reverence" for animals and nature. As one scholar put it, "Caring about other beings, about life, about our planet is aboriginal to our humanness."[52]

For example, the scent of soil evokes our profound connection with it; if you've forgotten what soil smells like, go outside and trowel up some. Botanist and Potawatomie Nation member Robin Wall Kimmerer writes: "Recent research has shown that the smell of humus exerts a physical effect on humans. Breathing in the scent of Mother Earth stimulates the release of the hormone oxytocin, the same chemical that promotes bonding between mother and child."[53] We are all, on some level, in love with the living world.

Yet, in the last 200 years human interactions with the living world changed dramatically, as people spent more time enclosed in places like cars, homes, and workplaces. Now, 51 percent of Americans spends *no* time outside on a normal day (excluding walking from house to car to office).[54] Another 30 percent spent less than an hour outside each day. In an increasingly urbanized world, moments of connection with the living world are even more crucial—not just to understand and experience living nature, but also for health. Just two hours a week interacting with nature—in a city park or other green space—increases individuals' satisfaction and health.[55] Interacting with nature results in improved mental and spiritual health, decreased anxiety, and improved concentration.[56] For kids, nature is extremely important for childhood development and health, and animal interactions have helped some children with autism spectrum disorders.[57]

Interbeing and Interrelationship

Since ecocentric beliefs may be foreign for some people, how do you acquire this orientation and way of interrelating? Think of it as a series of shifts that expand your view and thoughts outward beyond humans, and that notice humans in constant interaction with all that is not human. As Albert Einstein wrote above, "widening our circle of compassion to embrace all living creatures and the whole of nature."

In *Climate: A New Story*, Charles Eisenstein labeled this shift from a story of separation to a new story of *interbeing*, which recognizes that "existence is relational":

> Who I am depends on who you are. The world is part of me, just as I am part of it. What happens to the world is in some way happening to me. The state of the cultural climate or political climate affects the conditions of the geo-climate. When one thing changes, everything else must change too. The qualities of the self (sentience, agency, purpose, an experience of being) are not confined to humans alone. And the results of our interactions will come back to affect ourselves, inescapably. Interbeing . . . must be a way of seeing, a way of being, a strategic principle, and most of all a felt reality.[58]

First, consider all the ways that your physical body is intimately interdependent with the living world. Each day you take 23,000 breaths and you drink quarts of water (and your body is 60 percent water).[59] Your body is nourished with plants (or animals who ate plants), plants who were pollinated by insects and rooted in soil and fed with sunshine and water. On your body are clothes from plant fibers (cotton, flax, bamboo, hemp) or from animal fibers (wool, silk, angora); synthetic fibers like polyester, rayon, nylon, and fleece are made from petroleum products (aka fossilized sunshine). Each day, your body's OS (operating system) is the living world.

Now, shift your attention to all that is alive, all around you. Our pets and wilder animals possess intelligence and emotion; they are active beings with agency and free will who interact with you. Think of the carrot who you just ate as being alive—once rooted and growing in the dark soil and now providing you energy and interacting with your stomach.

It might be harder to think of your T-shirt or boulders along a path as "alive." They seem passive and inert. Yet, such things are just as vital and vibrant as animate beings, argues Jane Bennett in *Vibrant Matter*.[60] If you think of *vitality* as the capacity of things to affect you and act as forces with their own unique propensities and tendencies, that boulder is truly vibrant. It affected you to walk around it, it provided a sitting spot, and perhaps you thought it handsome. All kinds of "matter" produce effects and alter how things happen just by their presence—carrots, rocks, rainstorms. Humans are embedded in a tangled web of acting and being acted upon. For Bennett, the fantasy that we humans are somehow in charge of all those vital things prevents us from seeing our connection to all matter.[61] Your T-shirt is not alive in the way that the cotton plant was respiring and growing, but it is wholly made from earth elements and it's vibrant in its capacity to affect you. What you might think of as "dead" nature—a frying pan, a pencil—are things of the earth with a degree of vitality if not animacy.

Another important shift toward interrelationship is seeing other creatures and elements as active *subjects* with intrinsic value. When you walk among trees, what do you see? You feel the shade or see colorful leaves, or hear leaves or needles rustling. You know trees are alive, but chances are that you perceive trees as passive *objects* of your gaze and consideration. How might you flip this view? Think of them as active *subjects*, autonomous in their ability to direct and live their own tree-lives (wholly without you). They make choices and respond to Others in their surroundings (fungi, rain, insects, and mammals). They even communicate with other trees.

Increasingly, researchers think of trees not as individual plants but as communities with extraordinary interconnections and communication. Some scholars consider a forest as a superorganism, where different species of trees "cooperate" and "share."[62] Above ground, trees release scent signals (as a gas or chemical compound) to other trees to warn of insect attacks. Below ground, trees are connected by a dense network of mutualistic fungi that operate like fiber-optic Internet cables. Under a single footstep are hundreds of miles of a fungal network that links the tree roots, sending nutrients (like nitrogen, phosphorus, and water) back and forth as well as chemical and electrical signals. Dr. Suzanne Simard of the University of British Columbia coined the term "wood wide web" for the way that trees exchange news about insects, drought, and other dangers.[63] Through this network, trees are able to send nutrients to fellow trees who might be struggling under stress—operating more like a community than individuals competing against one another.

Plants are also active subjects. They monitor the visible environment all the time, writes biologist Daniel Camovitz in *What a Plant Knows*: "Plants see if you come near them; they know when you stand over them. They even know if you're wearing a blue or a red shirt. They know if you've painted your house or if you've moved their pots from one side of the living room to the other."[64] Plants "see" in the sense that light stimuli are received, interpreted, and in some manner, reacted to. "Eyes" in their seedling tips have phytochrome receptors, a light-activated switch that senses light and determines its color. While we see daylight as colorless, plants see light in a rainbow spectrum of differing wavelengths, using blue light to know which direction to bend and red light to measure the length of nighttime.

When you view a plant or tree as possessing vitality, agency, and intrinsic value, you might be more aware that you (and Others) exchange the same air and soak in the same sunshine. Your awareness might suggest different paths of action, such as not harming the plant (and bees and animals who visit it) with chemicals. At a macro level, the path might involve restoring trees or damaged lands or granting them equal "rights' (more on that in a moment). Think of how these shifts affect language and communication. Plants and creatures become fellow living beings and not "its" to control and "manage" in human-centered ways. You would communicate that what harms the living world also harms you and your fellow humans—that we are all "sensitive groups."

Lessons from Indigenous Peoples

In the previous chapter, you learned that indigenous people generally believe that the entire world is alive and humans hold a reciprocal relationship with all of life. Here, we'll look at the lessons to be learned from their ecocentric beliefs, particularly in light of climate change.

Around the world, indigenous people live according to the rhythms of nature, which is the source of their identity and sustenance. Indigenous peoples worldwide now face enormous pressures on their traditional lands (or have been forcibly removed from them). They experience encroachment and development and have suffered great losses of traditional indigenous knowledge. Climate change has brought profound changes to their lands and lifestyles.

In indigenous communities, cross-generational values help younger generations become part of a larger interconnected world and face ecological changes to traditional practices. In a study that compared two indigenous agricultural communities (one in New Mexico and one in Peru), seven key values transmitted a sense of interdependency with the living world: thankfulness and acknowledgment of all things, humility, nurturing and suffering, respect and love, togetherness, acting with faith, and sharing. [65]

Two essential indigenous lessons are gratitude and reciprocity. Kimmerer refers to plants as skilled and generous beings: "plants know how to make food and medicine from light and water, and then they give it away."[66] Kimmerer also explains that many Indians begin each day with an allegiance of gratitude to all members of the living world, sending greetings and thanks to waters, fish, other people, birds, trees, animals, the four winds, moon, stars, and sun.[67] Words of thanks treat Others as kin, neighbors, and fellow sovereign beings—not passive objects. With gifts freely given comes responsibility and reciprocity. You take care of the land and it will take care of you; you take something, and you must give something in return.

Recently, I was raking leaves from the towering silver maple in my backyard. It was the fourth time I'd raked that month, and I was impatient to move on to other weekend tasks. Then I caught myself: rather than viewing this as a chore to be done with, I should express my gratitude to this huge tree and all that it freely gives: it shades my house and yard from intense summer sun, its leaves are beautiful, birds build nests there, and my swing hangs from a large branch. It needs so little from me and gives so very much; it is indeed a constant and beautiful being. I am not a Native American, but the concepts of gifts and reciprocity have great wisdom for me.

Gratitude and reciprocity are also evident in language. Kimmerer points out that in modern Western thought, "land" is conceptualized as property, capital, and providing "natural resources" and "ecosystem services." However, to indigenous people, "land" is perceived as identity, nonhuman relatives, connection to ancestors, library, pharmacy, healer, sustainer, home, and moral responsibility.[68]

The Rights of Nature Movement

In most cultures (even more ecocentric ones) responsibility and care-taking of Others and ecosystems doesn't always occur. Some indigenous tribes, cities, and countries have taken steps to ensure that all beings are considered sovereign and are protected by granting legal rights. This is not a new idea. In 1972, Christopher Stone published *Should Trees Have Standing?*, a discussion of the legal standing of nature in a court of law.[69] His thesis was that the voiceless elements of nature should be granted legal rights, and it launched a worldwide conversation.

Almost forty years later, a burgeoning Rights of Nature movement seeks nature rights, akin to civil rights or human rights: nature has the right to exist, thrive, evolve, and be restored. Through its guardians, nature can defend and enforce these rights against threats from humans. Laws of a nation or tribe are an expression of its values and beliefs,[70] including toward Others.

A special election in 2019 in Toledo, Ohio, granted "rights" to Lake Erie: "Lake Erie, and the Lake Erie watershed, possess the right to exist, flourish, and naturally evolve. The Lake Erie Ecosystem shall include all natural water features, communities of organisms, soil as well as terrestrial and aquatic sub-ecosystems that are part of Lake Erie and its watershed."[71] This decision rejected conventional ways of relating to nature—as property that is ours for the taking, as an object rather than a subject. It also rejected a limited economic or utilitarian valuation of nature. Instead, humans see Lake Erie as a complex and interconnected ecosystem, and citizens can sue on the lake's behalf if it's being harmed.

Three scientists argued in *Science* that legal rights for nature could protect natural systems in ways that existing laws have not.[72] For example, the federal Endangered Species Act works to protect a species (and its habitat) but it does not grant it a right to exist. Animal rights and human rights prioritize the individual animal or person, but rights for nature encompass ecosystems and rivers, and acknowledge the intrinsic value of their complex communities. The scientists wrote that rights-of-nature advocates maintain that "environmental devastation is a moral wrong that ought to be stopped. This claim is not grounded in scientific evidence but is no less valid than the assertion that harming humans is a moral wrong."[73]

In 2008, Ecuador became the first country to recognize the rights of nature in its constitution, thanks in large part to indigenous activists wanting to protect Pacha Mama, an indigenous earth goddess: nature had "the right to exist, persist, maintain, and regenerate vital cycles, structure, functions, and its processes in evolution."[74] In 2014, New Zealand recognized the rights of the Te Urewere forest as part of a treaty settlement with the Maori tribes. Three years later, New Zealand recognized the legal "person" Te Awa Tupua as an "indivisible and living whole, comprising the Whanganui River from the mountains to the sea." In 2017, Colombia granted rights to the Rio Atrato River, and India recognized the Ganges and Yamuna rivers as legal persons. Colombia's highest court ruled in 2018 that the Colombian Amazon had the same legal rights as a human. Nature rights laws have been enacted by more than thirty US municipalities, including Pittsburgh and Santa Monica, as well as by several US indigenous tribes. In 2019, the Yurok Tribe in Oregon declared rights of personhood to the Klamath River.[75]

Does granting rights to nature effectively protect it? So far, attempts to defend the rights of nature have had limited results, and countries like Ecuador and Bolivia have not been able to slow environmental degradation.[76] In addition to internal politics and conflicts with corporate or human rights, there are questions about how to define the rights-bearer and exactly how nature may claim its rights (through human "guardians"). Yet several rights-of-nature designations have had success in the courts, including one in Ecuador where a river was being polluted by a construction project. People successfully sued on the river's behalf.

Even if the rights of nature movement has not achieved immediate or far-reaching change, it has shifted anthropocentric discourse in a critical way. The law compels people to now pause and think of Lake Erie as a whole, living being. Labeling a river or a country as a "legal person" doesn't personify as much as it grants being-hood to subjects with vitality and intrinsic value.

A TRANSITION TO A NEW RELATIONSHIP

A "citizen" is typically viewed as an inhabitant of a state or nation, but I've argued that what we really inhabit is planet Eairth. Practicing your *Eairth Citizenship* is not about undertaking a particular behavior, but about shifting your way of being in the world to include full consideration of it. Here are four suggestions to get you started.

First, *learn about the living world* around you. Where does your water come from, where does your garbage go? How much precipitation do you get? What's the name of the magnificent tree across the street and the black-and-yellow bird in it? How is the living world discussed and valued by those around you?

Second, *be in the world*. Examine your beliefs about the living world, and what you might shift to be an Eairth citizen. Describe your biophilia—what do you love and treasure in the living world and what gives you joy? Prescribe for yourself at least two hours a week outdoors.

Third, *be grateful for the world*. Fill your days with gratitude for the Other. I really enjoy saying this grace from Buddhist teacher Thich Nhat Hahn: "In this food, I clearly see the presence of the entire cosmos supporting my existence."

And finally, *give back—especially by not taking*. It's admirable to help restore environmental damage; it's more important to prevent it. In every decision—personal, cultural, political—fully consider Others' rights to live and prosper. Choose not to diminish Others' homes and not to poison. Build highway crossings to allow safe passage. In daily practice, prevent the need to log more, clear more, use more (and thus help prevent the next pandemic).

Although beliefs are held by individuals, a dominant belief system such as anthropocentrism greatly inculcates and reinforces how a culture operates. Several centuries of humans' rush to build our cities and structure our lives around fossil fuel energy was undertaken without realizing how those fuels would ravage the climate; initially the climate crisis was "by accident." However, the "accident" was driven by anthropocentric beliefs that our lives were separate from the nature we thought we mastered.

Some have suggested that the climate-driven transition before us is potentially as profound as the Axial Age, a period when the world seemed to change quickly and with much uncertainty.[77] From the third to the eighth centuries B.C., five distinct civilizations underwent huge transformations that included literacy, iron-working, urbanization, and market economies. It was a period of enormous collapse from which emerged new ways of seeing and being and new beliefs and practices. Some wonder if we are now entering a second Axial Age[78] for the climate crisis will continue to profoundly alter life on this planet for all beings.

Ecophilosopher Joanna Macy labeled the current climate-driven transition the Great Turning,[79] which she believes will be of comparable scope and magnitude to the Agricultural Revolution five thousand years ago and the Industrial Revolution 300 years ago. Macy says the Great Turning transformation will reinvent the whole basis of society and our relationships with one another and with Earth. She calls the Great Turning "the essential adventure of our time":

> In the early stages of major transitions, the initial activity might seem to exist only at the fringes. Yet when their time comes, ideas and behaviors become contagious: the more people pass on inspiring perspectives, the more these perspectives catch on. At a certain point, the balance tips and we reach critical mass. Viewpoints and practices that were once on the margins become the new mainstream.[80]

One dimension of the Great Turning is a shift in consciousness, such as toward ecocentrism and interbeing: reciprocal relationships where all humans and the living world are intrinsically valued. This shift would make very clear what is of *primary* importance (a healthy living world that is the source of everything) versus what is *secondary* (social constructs such as the marketplace and societal institutions).[81] If anthropocentric consciousness does *not* shift in the face of climate change, the result would likely be greed and selfishness amidst conflict and scarcity.

Shifted consciousness can also emerge from religious moral teachings. Climate activist (and former Archbishop of Canterbury) Rowan Williams said that Christians are told to love our neighbors as ourselves. He said that when Jesus was invited to define who counted as a "neighbor," he replied, it was "the one who gives you life."[82] In that sense, "neighbors" populate the entire living world—in Others, in air and water and soil. Instead of withdrawing in the face of the climate change challenge, Williams says, "We need a fresh sense of the delight to be found in human and non-human creation alike, a fresh sense of the importance of living in *attunement* with who we are and what the world is."

Many people feel that attunement with the living world when they spend significant time outside—ranching, hiking, science field work, farming, birding. You read clouds and weather, recognize sounds and smells, and feel extremely comfortable and "at home." Your actions and decisions are made with an awareness and knowledge of Others and the larger landscape.

Even if you lack such experiences, you might have felt this sense of alignment in personal relationships, or even in a workplace. You developed relationships that were mutually supportive, fulfilling, benefitting. You recognized when adjustments and compromises to your behaviors were needed that were beneficial to all parties. You were aware of not wanting to harm others and allowing others to prosper. Now, take your knowledge of the qualities of reciprocal human relationships and stretch and adapt that to your relationship with the living world.

Alignment recognizes that there are simply healthy and unhealthy ways of living on this planet. Williams said, "There are ways of learning to live better, to make peace with the world [that] will limit the disease and destruction; they may even be seeds for a future we can't imagine."[83]

Here's one final thought about how to induce a shift in consciousness. For Charles Dickens' character Scrooge, the shift "was through a confrontation with beauty, suffering, and mortality. It was through a connection to what is real. One might call it an *initiatory experience*."[84]

Perhaps you or someone you know has had an initiatory experience with climate change—when you felt connected to what is real and happening and were confronted with the suffering and the beauty, the mortality and the life. You witnessed vanishing places where you live or visit, or experienced ill health from the environment. Information can sometimes prompt such an experience—maybe you watched a shocking documentary or did a close reading of a religious text about creation care. Maybe you connected the dots between human actions that returned as grave consequences, such as the journey of plastic to ocean to halibut and back to you. An initiatory experience often emerges from direct, grounded experience in a place that you love and with people you love.

ALL FLOURISHING IS MUTUAL

After the environmental social movement's activism in the 1960s and 1970s, some people thought our beliefs about the environment had truly changed. After all, the US had new and important laws regulating clean air, clean water, and wilderness. President Jimmy Carter put solar panels on the White House. (Then, President Ronald Reagan removed the panels, which have gone up and down several times since.) The new environmental laws changed some material conditions: rivers no longer caught on fire, the ban on leaded gasoline helped clear skies, and DDT was no longer manufactured in the US. However, in the last fifty years, environmental regulations have been subject to shifting political winds and significant rollbacks under various presidents.

This history demonstrates that addressing climate change and related ecological crises requires not just new laws and regulations, but also a significant shift from anthropocentric beliefs to a more connected way of living in the world. Given the extraordinary interdependence of all life on Eairth, all flourishing is mutual. Imagine, as Albert Einstein suggested, that you clearly see the optical delusion that you are separate from the living world. Imagine widening your circle of compassion to embrace all creatures and the whole of nature. Then imagine yourself as an Eairth Citizen taking your beliefs and actions out into the larger living world.

NOTES

1. Vivian Li, "Marine Plastic Pollution Hides a Neurological Toxicant in Our Food," *Phys.Org*, September 6, 2019, https://phys.org/news/2019-09-marine-plastic-pollution-neurological-toxin.html.

2. https://seagrant.unh.edu/project/research/role-plastic-marine-debris-magnification-methyl-mercury-seafood.

3. A quote from a letter of consolation that Albert Einstein wrote to a grieving father named Robert S. Marcus, political director of the World Jewish Congress, whose young son had just died of polio. The letter was later included in *The Quantum and the Lotus: A Journey to the Frontiers Where Science and Buddhism Meet* by Matthieu Ricard, 2004, Broadway Books.

4. Peter Gleick and Heather Cooley, "Energy Implications of Bottled Water," *Environmental Research Letters* 4 (2009), doi:10.1088/1748-9326/4/1/014009.

5. Christopher Joyce, "Plastic Has A Big Carbon Footprint – But That Isn't the Whole Story," *NPR.Org*, July 9, 2019, https://www.npr.org/2019/07/09/735848489/plastic-has-a-big-carbon-footprint-but-that-isnt-the-whole-story.

6. Rebecca Harrington, "By 2050, the Oceans Could Have More Plastic than Fish," *Business Insider*, January 26, 2017, https://www.businessinsider.com/plastic-in-ocean-outweighs-fish-evidence-report-2017-1.

7. Joyce, "Plastic Has a Big Carbon Footprint."

8. Julia Corbett, *Out of the Woods: Seeing Nature in the Everyday* (Reno: University of Nevada Press, 2018), 45.

9. "Recycling" plastic is highly problematic: only 7–8 percent of plastic in the past seventy years was recycled. Recycling is generally voluntary and many cities or countries lack recycling services. The most readily recycled plastics are #1 and #2, though five other plastic types carry a recycling number. Recycled plastic always degrades in quality—down-cycles—and remaking it requires more energy, water, and materials. Promising to recycle water bottles does not change their highly toxic manufacturing process or the fossil fuels from which they are made.

10. Rebecca Sutton, Diana Lin, Meg Sedlak, et al., *Understanding Microplastic Levels, Pathways, and Transport in the San Francisco Bay Region* (Richmond, CA: San Francisco Estuary Institute, 2019).

11. Both plastic and rubber are polymers. Synthetic rubber is made from crude oil, though it is made to be more elastic than plastic.

12. Stephen Leahy, "Microplastics Are Raining down from the Sky," *National Geographic*, April 15, 2019, https://www.nationalgeographic.com/environment/2019/04/microplastics-pollution-falls-from-air-even-mountains/.

13. Kevin Loria, "Think You Don't Eat Plastic?," *Consumer Reports*, June 2020, 30.

14. Luís Gabriel Antão Barboza, Luís Russo Vieira, Vasco Branco, Cristina Carvalho, and Lúcia Guilhermino, "Microplastics Increase Mercury Bioconcentration in Gills and Bioaccumulation in the Liver, and Cause Oxidative Stress and Damage in Dicentrarchus Labrax Juveniles," *Scientific Reports* 8, no. 1 (October 23, 2018): 1–9, doi:10.1038/s41598-018-34125-z.

15. Rosanna Xia, "The Biggest Likely Source of Microplastics in California Coastal Waters? Our Car Tires," *Los Angeles Times*, October 2, 2019, https://www.latimes.com/environment/story/2019-10-02/california-microplastics-ocean-study.

16. Chelsea M. Rochman, Akbar Tahir, Susan L. Williams, Dolores V. Baxa, Rosalyn Lam, Jeffrey T. Miller, Foo-Ching Teh, Shinta Werorilangi, and Swee J. Teh, "Anthropogenic Debris in Seafood: Plastic Debris and Fibers from Textiles in Fish and Bivalves Sold for Human Consumption," *Scientific Reports* 5, no. 1 (September 24, 2015): 1–10, doi:10.1038/srep14340.

17. Rebecca Dzombak, "There's No Corner of the Globe Safe from Microplastic Pollution," *Massive Science*, June 10, 2019, https://massivesci.com/articles/microplastic-pollution-pyrenees-mountains-soil-air-environment-health/.

18. Marlowe Hood, "Do You Consume a Credit Card's Worth of Plastic Every Week?," *Phys.Org*, June 12, 2019, https://phys.org/news/2019-06-consume-credit-card-worth-plastic.html.

19. Harvard Health Publishing, "Microwaving Food in Plastic: Dangerous or Not?," Harvard Medical School, *Harvard Health*, September 20, 2017, https://www.health.harvard.edu/staying-healthy/microwaving-food-in-plastic-dangerous-or-not.

20. Li, "Marine Plastic Pollution Hides."

21. Barboza et al., "Microplastics Increase Mercury Bioconcentration."

22. Courtney Howard, "Healthy Planet, Healthy People," *TEDxMontrealWomen*, 2018, https://www.youtube.com/watch?v=FgIYaklWOK4.

23. William D. Nordhaus, "Resources as a Constraint on Growth," *The American Economic Review* 64, no. 2 (1974): 22–26.

24. See www.footprintnetwork.org and https://www.fastcompany.com/90381219/weve-officially-exhausted-the-earths-natural-resources-for-the-year.

25. Jing Lin and Rebecca L. Oxford, *Transformative EcoEducation for Human and Planetary Survival* (Charlotte, NC: Information Age Publishing, 2011).

26. Jonathan F. P. Rose, "A Transformational Ecology," in *Moral Ground: Ethical Action for a Planet in Peril*, ed. Kathleen Dean Moore and Michael P Nelson (San Antonio, TX: Trinity University Press, 2010), 207–10, 207.

27. Caroline Lucas, "A Political View," in *This Is Not a Drill: An Extinction Rebellion Handbook* (London: Penguin Books, 2019), 141–45, p. 141.

28. Rose, "A Transformational Ecology," 208.

29. Bob Tostevin, *The Promethean Illusion: The Western Belief in Human Mastery of Nature* (Jefferson, NC: McFarland, 2010).

30. David Orwell, *The Future of Everything: The Science of Prediction* (New York, NY: Basic Books, 2008).

31. M. Cappellari, Gianni Turcato, Massimo Zannoni, et al., "Association between Short- and Medium-Term Air Pollution Exposure and Risk of Mortality after Intravenous Thrombolysis for Stroke," *Journal of Thrombolysis* 45 (November 28, 2017): 293–99; L. Panis, Eline B. Provost, Bianca Cox, et al., "Short-Term Air Pollution Exposure Decreases Lung Function: A Repeated Measures Study in Healthy Adults," *Environmental Health* 16 (June 14, 2017): 60; A. J. Goodrich, Heather E. Volk, Daniel J. Tancredi, et al., "Joint Effects of Prenatal Air Pollutant Exposure and Maternal Folic Acid Supplementation on Risk of Autism Spectrum Disorder," *Autism Research* 11 (November 9, 2017): 69–80; C. S. Malley, Johan C. I. Kuylenstierna, Harry W. Vallack, et al., "Preterm Birth Associated with Maternal Fine Particulate Matter Exposure: A Global, Regional and National Assessment," *Environmental International* 101 (2017): 173–182.

32. Amitav Ghosh, *The Great Derangement: Climate Change and the Unthinkable* (Chicago, IL: University of Chicago Press, 2016), 30–31.

33. Kenneth V. Rosenberg, Adriaan M. Dokter, Peter J. Blancher, et al., "Decline of the North American Avifauna," *Science* 366, no. 6461 (2019): 120–24.

34. Jennifer Bennett, "Ocean Acidification," *Smithsonian Ocean*, 2019, http://ocean.si.edu/ocean-life/invertebrates/ocean-acidification.

35. Remember, however, that manufacturing solar panels and wind turbines requires energy and new materials, and large installations have serious environmental costs. The first and best step: *drastically reduce* energy use, no matter the source. For more information: https://www.ucsusa.org/resources/environmental-impacts-solar-power and https://www.theguardian.com/sustainable-business/solar-power-parks-impact-environment-soil-plants-climate.

36. Dougald Hine, "Negotiating Surrender," in *This Is Not a Drill: An Extinction Rebellion Handbook* (London, UK: Penguin Books, 2019), 81–87, 85.

37. Karen Brulliard, "The Next Pandemic Is Already Coming, Unless Humans Change How We Interact with Wildlife, Scientists Say," *The Washington Post*, April 3, 2020.

38. Nick Paton Walsh and Vasco Cotovio, "Bats Are Not to Blame for Coronavirus. Humans Are," *CNN*, March 20, 2020, https://www.cnn.com/2020/03/19/health/coronavirus-human-actions-intl/index.html.

39. Thais Borges and Sue Branford, "Rapid Deforestation of Brazilian Amazon Could Bring Next Pandemic: Experts," *Mongabay Environmental News*, April 15, 2020, https://news.mongabay.com/2020/04/rapid-deforestation-of-brazilian-amazon-could-bring-next-pandemic-experts/.

40. A. Leiserowitz, E. Maibach, S. Rosenthal, J. Kotcher, P. Bergquist, M. Ballew, M. Goldberg, and A. Gustafson, *Climate Change in the American Mind: November 2019* (New Haven, CT: Yale Program on Climate Communication and George Mason University, 2019).

41. https://www.cbsnews.com/news/leonardo-dicaprio-hbo-climate-change-documentary-trailer-released-today-2019-05-22/.

42. Charles Eisenstein, *Climate: A New Story* (Berkeley, CA: North Atlantic Books, 2018), 56.

43. Rowan Williams, "Afterword," in *This Is Not a Drill: An Extinction Rebellion Handbook* (London, UK: Penguin Books, 2019), 181–84, p. 181.

44. Eisenstein, *Climate: A New Story*.

45. Julia B. Corbett, *Communicating Nature: How We Create and Understand Environmental Messages* (Washington, DC: Island Press, 2006), 27.

46. J. J. Piccolo, H. Washington, H. Kopnina, and B. Taylor, "Why Conservation Scientists Should Re-Embrace Their Ecocentric Roots," *Conservation Biology* 32, no. 4 (2018): 959–61, p. 959.

47. Aldo Leopold, *A Sand County Almanac* (New York, NY: Oxford University Press, 1949).

48. P. Ehrlich and H. Mooney, "Extinction, Substitution, and Ecosystem Services," *BioScience* 33 (1983): 246–54.

49. Piccolo et al., "Why Conservation Scientists Should Re-Embrace Their Ecocentric Roots," 960.

50. MEA, *Millennium Ecosystem Assessment: Ecosystems and Human Well-Being* (Washington, DC: Synthesis Island Press, 2005).

51. Stephen R. Kellert and Edward O. Wilson, *The Biophilia Hypothesis* (Washington, DC: Island Press, 1993).

52. Eisenstein, *Climate: A New Story*, 145.

53. Robin Wall Kimmerer, *Braiding Sweetgrass: Indigenous Wisdom, Scientific Knowledge and the Teachings of Plants* (Minneapolis, MN: Milkweed Editions, 2013), 236.

54. Corbett, *Out of the Woods*, 9.

55. Mathew P. White, Ian Alcock, James Grellier, et al., "Spending at Least 120 Minutes a Week in Nature Is Associated with Good Health and Wellbeing," *Scientific Reports* 9, no. 1 (June 13, 2019): 1–11, doi:10.1038/s41598-019-44097-3.

56. Kara Rogers, "Biophilia Hypothesis | Description, Nature, & Human Behavior," *Encyclopedia Britannica*, https://www.britannica.com/science/biophilia-hypothesis.

57. Peter H. Kahn, *Children and Nature: Psychological, Sociocultural, and Evolutionary Investigations* (Cambridge, MA: MIT, 2002).

58. Eisenstein, *Climate: A New Story*, 9.
59. Corbett, *Out of the Woods*, 66.
60. Jane Bennett, *Vibrant Matter: A Political Ecology of Things* (Durham, NC: Duke University Press, 2010).
61. Bennett, *Vibrant Matter*, 7.
62. Peter Wohlleben, *The Hidden Life of Trees* (Vancouver and Berkeley, CA: Greystone Books, 2018).
63. Ann Strainchamps, "The Secret Language of Trees," *To The Best Of Our Knowledge*, April 28, 2018, https://www.ttbook.org/show/secret-language-trees.
64. Daniel Chamovitz, *What a Plant Knows: A Field Guide to the Senses* (New York, NY: Farrar, Straus and Giroux, 2017).
65. Elizabeth Sumida Huaman, "(S)He Who Will Transform the Universe: Ecological Lessons in Community Education from the Indigenous Americas," in *Transformative EcoEducation for Human and Planetary Survival*, ed. Jing Lin and Rebecca L. Oxford (Charlotte, NC: Information Age Publishing, 2011), 221–38.
66. Kimmerer, *Braiding Sweetgrass*, 10.
67. Kimmerer, *Braiding Sweetgrass*, 107.
68. Personal remarks by Robin Wall Kimmerer in a talk at the downtown *Salt Lake Public Library*, October 19, 2019.
69. Christopher D. Stone, *Should Trees Have Standing?: Toward Legal Rights for Natural Objects* (New York, NY: Avon Books, 1972).
70. https://therightsofnature.org/what-is-rights-of-nature/; Anna V. Smith, "The Kalamth River Now Has the Legal Rights of a Person," *High Country News*, October 28, 2019.
71. Sigal Samuel, "Lake Erie Now Has Legal Rights, Just like You," *Vox*, February 26, 2019, https://www.vox.com/future-perfect/2019/2/26/18241904/lake-erie-legal-rights-personhood-nature-environment-toledo-ohio.
72. Guillaume Chapron, Yaffa Epstein, and José Vicente López-Bao, "A Rights Revolution for Nature," *Science* 363, no. 6434 (March 29, 2019): 1392–93, doi:10.1126/science.aav5601.
73. Chapron et al., "A Rights Revolution for Nature," 1392.
74. https://therightsofnature.org/what-is-rights-of-nature/.
75. Smith, "The Kalamth River Now Has the Legal Rights of a Person."
76. Chapron et al., "A Rights Revolution for Nature."
77. Paul Kingsnorth, "The Axis and the Sycamore," *Orion*, February 2017.
78. Kingsnorth, "The Axis and the Sycamore."
79. Joanna Macy, Chris Johnstone, and C. S. E. Cooney, "Active Hope: How to Face the Mess We're in without Going Crazy," *Tantor Audio*, 2017.
80. https://www.activehope.info/contents.html/.
81. Derrick Jensen, "Forget Shorter Showers: Why Personal Change Does Not Equate Political Change," *Orion*, August 2009.
82. Williams, "Afterword," 183.
83. Williams, "Afterword," 182.
84. Eisenstein, *Climate: A New Story*, 29, emphasis mine.

Chapter 8

Telling a New Story

The assignment I gave them was: Tell the story of climate change up to the present time, beginning with "once upon a time."[1] Here is what Natalia Soleil wrote.

> *Once Upon a Time, there was the ocean and nothing else. Out of the ocean crawled life that eventually took many forms. After a passage of more time, one of the forms discovered fire, which gave a great advantage over the others. A many other thousand discoveries later, this form named itself and everything else through language and quickly overtook the planet, making a million more discoveries. However, despite how much the humans learned, there were innate qualities to this life form that made them an unfortunate choice to rule the globe, principally greed and arrogance. Humans, although they tried convincing themselves otherwise, were perpetually insatiable and selfish. These traits made them very short sighted and the earth and all forms of life suffered considerably. The majority of humans believed that the earth—and all on it—existed for their benefit. And due to their discoveries, they were able to exploit the earth—and all on it—to do just that. However, the earth and its life did not exist for a single entity, and thus the consequences of exploitation threatened humans' existence. Although this took another passage of time, the humans finally discovered this consequence and named it, although what to do about it is still being decided. It is important to note that whatever—if anything—the humans decide to do, the earth will continue to exist, because the humans are wrong: they need the earth and not the other way around.*

What Natalia and the other students wrote was the "old story" of the climate crisis that we continue to tell—and to live. Our decades-long inaction and inertia demonstrate that in many respects, we are fiercely attached to this "old story." It is a story of fossil fuel-driven consumption, technology, and continuing growth based on an illusion of separation and domination of the living world. Even in full view of global climate consequences, we have been unable (or unwilling) to rebuild the campfire and tell a "new story" that will lead us forward into a changed, positive future world. I get it: the old story is so comfortable and familiar; it's an old yarn that each one of us can easily tell. But this story is not helping us navigate uncharted waters and leave a vibrant world for those to come.

We use stories to find meaning and make sense of our lives amidst the larger world, and to link the past with the present and future. Right now, we are at an impasse between the old stories of fossil fuel culture and new ones yet to be written. Moving through this mythic story is akin to a rite of passage: separating from the old story and reckoning with the past, progressing through a transition that is both liminal and dangerous, and finally reincorporating fully into new ecologies and economies.

Stories can help direct how we move through this rite of passage. As Gary Nabhan writes in *Desert Legends*, to restore the land we must *re-story* it.[2] Stories link old and new paradigms and negotiate the blank pages with new experiences and ideas that lead to authoring new stories. Stories – told to each other in our communities and through literature, theater, music, film, art, and comedy – create emotional involvement that goes beyond awareness and knowledge. Fully engaging hearts, hands, and minds spawns the creativity and imagination needed to propel us into a new era. The use of stories is the final new direction that can help us communicate the climate crisis in absorbing and effective ways.

STORIES IN OUR LIVES

Storytelling doesn't just happen in Hollywood, TED Talks, and the Moth Radio Hour. We tell stories[3] every single day; as humans we have never not told stories. We are "wired for story"[4] and we dream in stories. Your stories begin with your own preconscious internal stories in response to a basic psychological imperative: you simply must reduce the enormous complexity of events and stimuli around you into something that makes sense and upon which you can act.[5]

Stories draw on events and relationships in the exterior landscape and project them onto your interior landscape. It's how you make the abstract concrete, organize your thoughts, relate to others, and express yourself.[6] Stories help you know who you are in the world and how you can live safely and happily in it. When asked, "How was your day?" you select a relatively small amount of potential material, and then shape it to fit into existing personal and cultural stories that make sense for you and for the person who asked the question. When you reduce a universe of complexity into a single story, that's a powerful construction of the perceptual world—including for the climate crisis.

We experience stories deeply in our bodies and brains, and they are powerful. If you've ever gotten lost in a story, you have experienced "narrative transportation," or the degree to which the plot activates your imagination through an empathic connection with the characters.[7] A good story can be a proxy for a vivid, personal experience[8]—you feel like you were actually there—which facilitates experiential (rather than analytical) learning. A compelling story can produce less counterarguing and resistance from the listener.[9] People transported into a story experience greater empathy and are more likely to show story-consistent beliefs and pro-social behavior.[10] Stories can influence cardiac activity, skin conductance, and emotional arousal.[11] Even negative emotions (fear and anxiety) presented in stories can compel action and urgency.[12]

Climate change stories based on lived experiences engage a variety of senses and allow the "invisibility" of climate change to be seen, felt, and imagined in the present and future. Stories can bring this phenomenon close to home and create space to grieve and experience loss, and also to reimagine a positive future. As we've seen, climate stories can encourage critical reflection of existing social structures and cultural norms, and facilitate collective engagement.

That doesn't mean that all the stories we tell ourselves are accurate, or even helpful at times. Imagine two people with vastly different views of how "safe" their city is—one person greatly fears burglars or walking alone, and one person never locks the car or the front door. They have told themselves very different stories about the same city, which significantly affects their own and others' behavior. Extend that analogy to climate change, and it's easy to see why climate change is urgent and serious in one person's stories, and distant and unproblematic in the stories of another. New compelling stories could engage both of them.

Stories teach us through their structure and function. At the end of a tale full of emotions, twists, and turns, the story is pulled together perhaps with a moral lesson or an outcome not imagined. You learn what things mean, what's important, and what to avoid. Stories operate through what they *are* (their structure) and what they *do* (their function, such as learning and aroused emotions). When you structure events in a certain order, include certain details, and shape what happens between characters, that conveys cause-and-effect relationships: because I did (or didn't) do X, then Y happened (or didn't happen). Stories are how we tell ourselves the significance of what happened.[13] This helps us discover the actions we must take in order to survive – whether or not they match what is "real" in the external world. Our internal stories can become self-fulfilling prophesies ("I locked the door, so no one broke in," or "I didn't lock the door, and no one came in.") Then, it's not just about what did or didn't happen, but about what *might* have happened. Climate change examples would be old stories that say we can't address the crisis because it might hurt the economy, or new stories that the future will be better than the present when we address climate change.

When a story (and its accompanying beliefs and behaviors) is repeated broadly and in myriad situations, it becomes a *mythic story* that seems to signify the nature of reality. Ken Baskin, a consultant who researches storytelling dynamics in workplaces, concludes, "once a story is accepted as mythic, it appears to produce knowledge that can override any contrary information in the external reality it describes."[14] That's a very powerful form of "knowledge." A mythic story greatly limits the range of behaviors and ideas, for "myth fixes behavior to what has worked, holding systems in those behaviors until they *must* change."[15] This contributes to a "senescent system"[16] that is inflexible and notoriously hard to change, even when behavior is increasingly self-destructive.

The old story of fossil fuel culture that Natalia told is a mythic story. She noted how the grand story of the climate crisis is an inflexible, entrenched, anthropocentric system even after we're aware of its danger to our survival. Think about the mythic stories you hear repeatedly: "Consumption is good for the economy. Shopping is the great American past time. Climate change is too expensive to fix. Fossil fuels make modern life wonderful. Climate change is normal and there's nothing we can do. The climate is changing, but we'll continue on this same path and just adapt."

The task before us is telling new stories that *replace* the harmful mythic ones that trap us in social inertia. We need new stories to envision a positive future, shape meaning, and create possibilities for new types of action. New stories must make the old mythic stories obsolete.

LEAVING FOSSIL FUELS AS A RITE OF PASSAGE

Nancy Menning, a scholar of environmental and religious studies at Ithaca College, says if we recognize the shift away from fossil fuel culture as a *rite of passage*, stories can better narrate the transition.[17] Various cultures hold ceremonies to guide progression through a rite of passage, such as entering marriage or transitioning from childhood to adulthood. Marking a rite of passage is important, including for climate change:

> [W]hen the passage from childhood to adulthood is ritualized, the community gathers with the individual to ensure a successful transition. . . . Rites of

passage structure social and biological transformation so as to minimize the personal and social disruption. . . . Viewed functionally, then, rites of passage focus attention, organize and coordinate collective action during dangerous and powerful transitions, and make manifest the fundamental interconnectedness of communities. Recognizing climate change as a rite of passage increases the likelihood of a successful passage to ecologies and economies of the future.[18]

A rite of passage is typically envisioned as a three-stage process: separation, transition, and reincorporation. The first stage is separating from a current or previous state, such as from the fossil fuel world when we become aware that we can no longer live in this old ecological era. Stories that stress a need for separation could describe the suffering brought to humans and Others by a changed climate.

The transition stage represents the passage—often spanning multiple generations—from the old era to a new one. Any big transformation is difficult, multifaceted, uncertain, and can feel as though change is taking "two steps forward and one step back." This is true for climate change; many embrace "renewable energy" and march for change, yet fossil fuel extraction and consumption continue apace as do the structural systems that support them.

At the final reincorporation stage, humanity is integrated into (and continues to create) a new ecological world. Celebrating and ritualizing a new ecological era might include an international holiday to celebrate fossil fuel independence, or arts festivals that highlight flourishing human and ecological communities.[19]

Menning applies the rites of passage to the powerful mythic story of the Bible's Exodus narrative. The Israelite people were in slavery in Egypt for many years. Their separation and movement toward a transition involved the emergence of a prophet (Moses), political struggles, and dramatic "ecological anomalies" such as their rapid escape through rising waters in the desert. The escaped slaves became "a new people in the desert" (a new cultural identity) though it was not without hardship. They missed the abundance of life in Egypt, and some members of that generation (including Moses) did not make it out of the desert. Their transition was one of "abundant power and stunning peril" and illustrates that an entire community undergoes a rite of passage; "individual and collective actions determine who will successfully complete the liminal period in the desert and enter into Canaan," the Promised Land.[20] When God later brought the community to the border of Canaan, the community feared the challenges yet before them; it would be their children who entered the Promised Land, the final reincorporation. As Menning concludes, the Exodus story corresponds to a liminal rite of passage for climate change:

> The details of the Exodus narrative—with its prophetic voices, political struggles, vacillations of commitment, and extended liminality—bring to mind corresponding details of the climate change story. The richness of the Exodus account also demonstrates the nested structure of rites of passage, in which the movements into and out of the liminal period . . . can themselves be seen as rites of passage. . . . Merely moving into a sustained awareness of climate change has been a rite passage in itself.[21]

Indeed, we have already experienced an insistent longing for the past, a sometimes overpowering fear of the future, and an enduring desire to engineer our way out of the crisis.[22] Previous chapters have noted the "nested structure" of transitions related to climate change: moving from social inertia to social change, moving in and out of grief and emotion, and the wide vacillation of public opinion and attention to climate change. How we choose to tell the climate story influences how we imagine the transition before us. Telling a story of apocalypse instills a narrative of danger and fear, which can make us feel paralyzed. Telling a story of climate justice orients a listener for ethical action and a hoped-for future. Telling a story of community action and rational hope models positive steps. It's important for new stories to harmonize our emotional, practical, and ethical responses in a way that overcomes the all too common disempowering tendencies of a transition.

The next sections discuss separating from the old climate story, navigating the transition period, and telling new stories that light the path toward reincorporation.

SEPARATING FROM THE OLD STORY

The storytelling exercise that began this chapter came from sustainability scholars Heidi Hendersson and Christine Wamsler. They discovered in their research that students' stories characterized the old story of climate change as one of technological dominance and human superiority over and separation from nature, which unfolds as ecological destruction.[23] Patterns within old stories included consumerism, scarcity thinking, a growth paradigm, and struggle. Not surprisingly, they found that anthropocentric beliefs are a dominant framework in the old story.

Because "we live in the wreckage of the old story,"[24] social inertia continues. We nibble around the edges of the old story but can't seem to make much headway. How can we begin to more fully separate from it? We've discussed some initial steps already. The alternative narratives of civil disobedience help disrupt the old stories, as do narratives that present the moral imperative of the climate crisis. These narratives expose problems of the old story and make the case for a celebration of interdependency with the living world. New messengers and narrators emerge to disrupt old climate change stories: athletes, environmental educators, indigenous peoples, teachers, and increasingly, politicians of various stripes. As we learned in the chapter about climate conversations, new stories help people connect the dots of how the old story brought us to this point, what the future portends unless we act, and then a reimagined future we truly want. It's important to thoroughly integrate emotions and the local reality of loved places in conversations and stories.

We are experiencing right now that separating from the old story and fully entering the transition is a "nested structure": we move in and out of separation from the old story, a fluid boundary we seem to cross a thousand times. The aftermath of another stalled hurricane seems to move us into transition, but then an economic slump moves us back to prop up the growth economy. A massive climate march conveys vital change, then an election changes that. And so it goes. Yet it is crucial to persist: continue to disrupt the old story and its failings and fill the void with new visionary stories. Over the last few years, the climate crisis has become far more present in daily life, and the rising urgency of the crisis calls for separating from the old story as quickly as possible.

THE LIMINAL TRANSITION

"Liminal" is an apt word to describe a difficult transition because it means occupying a position on both sides of a boundary or threshold in a space between "what was" and "what's next." It's a sensory threshold when "what's next" is only barely perceptible—like when darkness gives way to dawn and bit by bit, you see shadows, outlines, and shapes slowly emerge in the growing light. This liminal space is where all transformation takes place, if we learn to be comfortable with uncertainty. In your own life, you experience a liminal transition as a partner in a new relationship or as a new parent, moving back and forth across the boundary until you fully step into the new role.

Giving the uncertainties of the liminal threshold, it makes sense that a transition from fossil fuels is "the perfect moral storm"[25] with factors that threaten our ability to behave ethically: the global scope of the problem, its intergenerational scale, dispersion of cause and effect, skewed vulnerabilities, and institutional inadequacies. Stories during a liminal transition can emphasize the ethical predicaments but also the solutions. Stories can stress that climate injustice will worsen if institutions refuse action and current generations push the problem into the next generation. Stories can emphasize these real and present dangers to call us out of our apathy and perceived disempowerment, showing how climate change must draw together entire human and ecological communities and help us realize the shared nature of our flourishing.[26]

The transition, though long and arduous, is also a wide-open middle space with infinite possibilities. As any writer or storyteller knows, filling "blank pages" feels simultaneously terrifying and exhilarating. A terrifying possibility is that our transition is designed and dictated by just a few people of power and means, which would likely exacerbate suffering worldwide. On the exhilarating side, new stories can dream into ecologies and economies of the future and present them vividly and compellingly. Stories must work to invite and engage the largest and broadest possible collective to cocreate our transition.

Okay, this sounds like a very tall order, doesn't it. Maybe the blank pages seem someone else's to fill. Many people cannot envision a different, positive future world or see how their participation would matter. But consider this: millions of people around the world now possess a corporeal experience of what a cleaner and quieter living world would be like. During COVID-19 shutdowns, skies in Delhi, San Paulo, Bangkok, Bogotá, and thousands of cities changed from brown to blue, and water in Venice canals settled to reveal fish. Residents of Jalandhar, India saw the Dhauladhar mountain range for first time in thirty years. A reporter in Paris heard birds singing for the first time in years. Some locations experienced a 90 percent drop in noise. And seismologists found that the earth's surface was vibrating 30–50 percent less without fossil-fuel-powered machines rumbling across its surface. Millions of people experienced that the living world *can* be different; you cannot unsee a mountain range or unbreathe a clean breath.

It then becomes a matter of what to do with this knowledge and these memories. As writer and environmental activist Paul Kingsnorth reassured, at this point, "It is okay to be confused. It is okay to be small. It is okay not to know what to do. Really, the only thing that is not okay is turning away."[27] It helps to learn how to turn toward, instead.

Ecopsychologist Renee Lertzman reminds us that each person must choose to engage on his or her own terms and find an adequate "home" for contributing and participating creatively.[28] If a person doesn't feel she can contribute or make a difference, she is likely to withdraw emotionally from the climate crisis. A person's lack of engagement may reflect an inability to process loss or ambivalence. True concern might not be absent, but just seeking a home.

The concept of "contributing to" something meaningful is important so you don't feel as though your steps are insignificant. As discussed earlier, a key way to bolster significance is to make sure your actions focus beyond yourself toward larger systems and practices that need changing. Signing a petition or giving money is fine, but if you don't feel as though you are adequately "contributing to," keep looking. If someone is grieving the loss of forests from fires, he might "contribute to" an organization by planting trees or helping others create "defensible space" around their homes. Again, an excellent initial and meaningful way to contribute is through conversation and stories with all the important others in your life. Those conversations might provide meaningful insights into exactly how you can best "contribute to."

The back-and-forth of the liminal transition can be wearying. The science is full of bad news (and impacts are happening faster than predicted), the news is dreary, and so is the politics. So how can a person stay hopeful during the drawn-out liminal transition? As climate scientist and communicator Katharine Hayhoe expresses it, hope arrives with action:

> Hope doesn't come to me if I just sit there waiting for it to show up. Hope comes from action, from going out and looking for positive news and positive examples of what people are doing, looking for and engaging with people who share my values and concerns. That's incredibly hopeful, to be together in community.... Not giving up is actually what gives us hope.[29]

Since the biggest pop culture trend in 2019 was "climate anxiety,"[30] taking action and finding hope is powerful medicine.

It then becomes a matter of finding others who share your values and concerns and finding places where you can contribute your particular skills and passions. Maybe you are a great "connector," able to draw people together, organize, and make things happen. Maybe you are a physical guy who likes to work with his hands. Perhaps you can write, or teach, or fix machines and electronics. Maybe you're an artist, a playwright, a seamstress, a cook, or a supreme number-cruncher. Match those skills with the groups and organizations working to change systems and large-scale practices. Mobilize voters; register voters. If you are a health care worker, spread the news that our health and the health of the living world are one in the same. When you act and create community, you develop vast stores of hope that will sustain you through the liminal transition—and that simultaneously help shape it in powerful ways. In all these places, tell new stories of a positive climate future.

When people learn that consumer actions don't move the needle on carbon emissions, they ask me, "Okay, then what are things that do move the needle that I could get involved in?" Since two-thirds of carbon emissions come from electricity, cars, and buildings, focus on one of those sectors first; this is the approach of various states and cities to significantly reduce emissions. Perhaps you have good ideas for retrofitting and updating large buildings, or you think your town could begin its own electric co-op run by renewables.

In addition, here are five suggestions from Rebecca Willis, a professor of climate and energy policy at UK's Lancaster University, of actions to restrain carbon emissions right now.[31]

First, task all government departments to develop a climate plan so that they're obligated to show how they will contribute to *significant* emissions cuts.

Second, engage the public in developing climate strategies with citizen assemblies (discussed more below), and involve workers in designing a "just transition" from carbon-intensive industries like coal mining.

Third, enact "symbolic policies" to shift the "climate" for investing, which could catalyze radical change. For example, work to ban advertising for gas-powered cars (similar to the ban on tobacco ads), or work to allow people to generate and sell renewable energy at home.

Fourth, keep all remaining fossil fuels in the ground. Ban oil and gas exploration and end the $5.2 trillion spent on subsidizing the fossil fuel industry each year.

And fifth, Willis says don't invest time and energy on technologies such as carbon capture and storage. That method might play a future role in absorbing greenhouse gases, but it doesn't yet exist at a meaningful scale, and it distracts resources from cutting emissions *now*.

These five measures are being tackled by groups of all sizes all around the globe; see if you can connect with one and offer to help.

In addition, there are hundreds of ways to promote natural carbon mitigation through restoring land, shifting agricultural practices, planting trees, composting, eliminating food waste, and transforming monocultures to biodiverse habitats. You could initiate some of these in your own neighborhood or city.

TRULY ENGAGING WITH HEARTS, HANDS, AND MINDS

News and information about the climate crisis circulate every day, which contributes to awareness and cognition, but it has not led to sufficient *engagement*. When someone is engaged, he or she has a personal state of connection or involvement, particularly that includes emotions and behaviors.[32] In other words, a person is engaged with heart, hands, and mind[33] at a deeper and more interactive level. True engagement encourages people to be participants, stakeholders, and co-creators of their future.[34] Communication can stretch far beyond "flat" written messages about climate change and expand where and how engagement happens.

Engaging new stories can be told in a wide range of places and settings you might not have considered—from small groups to large collectives, from civic dialogue forums and the humanities to visual arts, literature, performance arts, and even comedy. In whatever form, communication that motivates engagement must be inviting, inclusive, and full of emotion, purpose, and meaning. When someone is truly engaged with climate change, she or he understands and cares about the problem, feels that things can be done to address it, and interacts with others to create avenues to accomplish change. Here are some inspiring examples of new stories around the world.

Social interaction touches people deeply and thus motivates their interest and sustains engagement. Sociologist Robert Brulle emphasizes the need for dialogic communication about climate change in the civic sphere to spur democratic engagement and public dialogue, and hence effect social change.[35] Public forums in libraries, theaters, classrooms, and parks bring people together to share their stories, questions, uncertainties, fears, and aspirations. Rather than confronting climate change in isolation, individuals are able to collectively invest in each other, share stories, and determine potential courses of action.

A type of civic engagement championed by Extinction Rebellion is citizen assemblies, in which representative citizens are chosen to discuss and negotiate a city's or country's separation from fossil fuels. The idea is that addressing a deep-rooted problem like climate change requires widespread public involvement, consent, and a constellation of national conversations about the liminal transition ahead. Key is that citizens' conversations and deliberations help redirect political inaction toward more productive decision making.

The country most actively engaged in citizen assemblies is the UK. As Climate Outreach reported, at the national Climate Assembly, 110 randomly selected but demographically representative people spent four weekends "talking climate" and coming to grips with how the UK could reach net-zero by 2050.[36] The national assembly is running in tandem with local gatherings, where cities form citizen assemblies and jury exercises to better understand the climate emergency policies that people want to see in their neighborhoods. Although assembly policies are not legally binding, citizen choices are an important redirection of political reality.

Beyond civic dialogue, the arts and humanities can effectively engage audiences in climate change by disrupting dominant worldviews and reaching beyond scientific data to help reboot the cultural conversation, particularly in ways that grapple with the difficult moral dimensions. As Danish artist Olafur Eliasson said, art can be a great communicator in civic society, especially because politicians and the private sector have let us down and are obsessed with the immediate and not the long term.[37]

Visual arts, dance, literature, theater, music, film, and comedy can engage us in five key ways: through storytelling, by creating corporeally sensed and felt experiences, illustrating our interdependency with the living world, engaging our emotions, and connecting us with place.[38] Here are examples of how different art forms have utilized these five elements to engage audiences with the climate crisis.

In addition to the performing arts and film, literature (fiction, nonfiction, poetry) is obviously story-based. Earlier, we discussed the unexpected arrival of monarch butterflies at Dellarobia's rural Tennessee farm in Barbara Kingsolver's *Flight Behavior*, which is an example of the new genre "cli-fi" or climate fiction (where a fair amount tends toward the apocalyptic or dystopian). Nathaniel Rich's *Odds Against Tomorrow* was a novel about a disastrous hurricane ravaging New York City—written *before* Hurricane Sandy hit. Rich never used the phrase "climate change"; he said the phenomenon was a foregone conclusion, but the real story is in how we deal with it.[39] Cli-fi literature can effectively engage the human heart and condition.

Just as you learned to compose your personal climate story, there are widespread storytelling projects and oral histories that encourage expression of one's personal engagement with climate change and environmental issues. The Climate Narrative Project, launched by author and activist Jeff Biggers, has presented Ecopolis performances from Appalachia to Silicon Valley. Local residents and actors use story, music, and art to consider green paths to regenerate their cities and regions. Said Biggers, "I tell every campus or town I visit to appoint a Climate Storyteller-in-Residence. Instead of hiring another bureaucrat to convene a task force . . . they should hire artists and writers and playwrights and farmers to put more regenerative ways of living, planning, and development on the center stage of our daily operations."[40]

The Climate Stories Project invites individuals to record personal oral histories to bring immediacy and individual faces to climate communication. A map of recorded stories shows dots across Africa, Europe, Australia, Japan, Canada, US, and the Philippines. Similarly, Climate Stories North Carolina captures in short videos the stories of farmers, beekeepers, fishermen, hunters, apple growers, and others who have been directly affected by changes in the climate.

*Changing Climate, Changing Health, Changing Storie*s was a digital storytelling project of Ashlee Cunsolo Willox and colleagues' work with the Inuit government in Rigolet, Labrador, Canada, where residents documented their experiences of how the changing climate was affecting local ecosystems.[41] Inuits told stories of how climate changes contributed to feelings of depression, anxiety, and fear, as their subsistence hunting and fishing became more difficult. These stories counter the tendency to view climate change as merely a theoretical threat.

I find myself seeking comedy to put climate news into perspective. For centuries, comedy has broached difficult topics and revealed contradictions, expressed emotions, and relieved tension. Birte Loschenkohl, a philosopher at the University of Essex, does not recommend that we turn away from tragic and apocalyptic narratives entirely because there is much truth and value to them. But she says we would do well to supplement them with comic reflections on our relationship with nature and our ability to act in the face of hopelessness. Loschenkohl says if we think of our own agency more like comic heroes (who don't control their environment yet manage to muddle through and make lots of mistakes) "this might help us persevere in view of the seemingly impossible tasks ahead."[42]

In addition to comedy in theaters and films, late-night TV comedians routinely discuss politics and climate change; a sketch by Jimmy Kimmel featured kids explaining climate change to President Trump.[43] The faith-based organization Operation Noah and even some NASA scientists have produced comedy shows. Studies have found that comedic content can positively influence viewers' perceptions about climate change[44] and that humor is useful for engaging the public, in particular prompting eighteen- to twenty-four-year-olds to become more politically engaged.[45]

Comedy was used as persuasive communication by the Environment Minister of Agriculture in Egypt.[46] An hour-long comedy theater performed in fifty villages in upper Egypt told the story of a farmer who refused to pay for a new, more water-efficient agricultural canal. Agriculture officials used this format to reach farmers with an enjoyable message that also taught how to better cope with climate changes.

In their songs and stories, musicians, actors, and film makers provide audiences with corporeally sensed and felt experiences and emotions.[47] Although Hollywood and traditional TV programming have largely neglected climate change, documentary films have somewhat filled the void, such as Matthew Testa's *The Human Element* (2019), which features not scientists but people affected by the changing planet: firefighters, anglers, and coal miners.

Visual arts and art installations have produced significant public engagement through sensory, emotional experiences. Olafur Eliasson's "Ice Watch" brought twenty-four icebergs from Greenland to the sidewalks of London, which halted and mesmerized passersby observing the meltwater quietly dripping into the Thames.[48] Another Eliasson installation was "The Weather Project" at the Tate Modern museum in London (2003–2004). Weather is a fundamental encounter with nature that individuals truly participate in. His exhibit was created literally out of smoke and mirrors. A semicircular bank of sodium yellow streetlights produced a huge, artificial indoor sun in the dark hall, where viewers were engulfed in the sickly, misty gloom (created with sugar smoke) of a hazy atmosphere. It was truly corporeal; viewers lay on floor for hours on end, drawn to the powerful, ominous experience.

As any consumer of visuals knows, the tens of thousands of photos of climate change are not equally powerful in evoking active and corporeal responses. As photographer Gary Braasch reflected, seeing is not necessarily believing when it comes to climate change.[49] Many symbolic images have become clichés (polar bears, glaciers, smokestacks) and reinforce that climate change feels distant from everyday life.[50] More empowering visuals show climate change effects close to home and tell visual stories of how ordinary people are harmed. One study did this through photo-elicitation, where research participants in Lakes Entrance, Australia, took photographs of meaningful places where they lived and that they thought would be threatened by flooding or sea-level rise.[51] Researchers interviewed the participants and discussed the photographs, revealing strong emotional connections to place. The citizens described the meaning the place had for them, their families, and community, causing them to reflect upon current climate mitigation policies. Here, the past, present, and future were brought together, as participants emotionally and corporeally felt the gravity of climate change.

Sensing climate change as a threat to human life is predicated on recognizing one's interdependency with the world, as well as the world's mutability. An empowering example of science and humanities joining together to instill knowledge of interdependency is the work of ecologist Nalini Nadkarni, who studies forest canopies. She brings artists and scientists together for "Canopy Confluences" in the Pacific Northwest where they climb trees, collect samples, and learn about interconnecting ecosystems.[52] Participants learn about the dynamic lives of trees and the important ecological functions they provide for all life on Eairth. Children from the inner city, who had previously never climbed trees, integrated their experiences and lessons into original hip-hop songs. Others painted murals of forest ecosystems in their cities to bring this awareness into urban landscapes.

An attachment to "place" refers to the emotional connections that individuals have with a physical environment, bonds that arise from familiarity and a sense of belonging.[53] We find great security and identity when we are surrounded by constant and familiar environments. Because of this, our attachment and bonds with places can be an important avenue to engage people and help them witness and understand climate changes to loved places, particularly local ones.

The possibilities for place-based engagement are infinite—outdoor environmental education for all ages, scientists and citizens touring critical watersheds or habitats, neighborhoods restoring green spaces and native species. The burgeoning "citizen science" movement sends individuals and families outside to locate milkweed plants for monarchs and report bird sightings. The millions of records provided by citizen scientists have documented how climate change has altered seasonal timing in virtually all ecosystems.

The Little River Band of Ottawa Indians created an inspiring restoration project that united the community around local sustainability concerns.[54] Tribal members came together to celebrate and reintroduce Lake Sturgeon back into the Great Lakes watershed, affirming their connections to each other and the larger environment.

As any communicator, activist, or environmental professional knows, moving people to engage with the climate crisis is difficult. Yet, engaging is crucial to successfully navigate the liminal transition to a new ecological era. The personal benefits are profound as well; interacting with social others outdoors, through the arts, or in civic dialogue are sure ways to reduce climate anxiety and bolster hope.

REINCORPORATION: ENVISIONING NEW STORIES

David Bowie said that "Tomorrow belongs to those who can hear it coming."[55] What he describes is an ability to intuit the future – to hear, see, and imagine what lies ahead. And that seems to be largely missing for us as individuals and in the communication of the climate crisis. Some say that we're in this crisis precisely because of our failure to imagine meaningful alternatives to where we are right now.[56] We are averse to solutions and to change because we have trouble picturing a changed future world in which we want to live.

Ed Miliband, UK member of Parliament, says enough of the nightmare scenarios about climate change—it's time to paint the dream. He believes that tackling the climate and ecological crisis urgently requires reimagining how we live and work and offering a compelling, attractive vision of the future—one that also tackles social and economic injustice: "Once we begin to confront the way our economy is configured, how we measure and produce value, the hours we work, the way we consume, a truly different world is possible."[57]

A key piece of hearing the future is imagining the future we *want* to create—not just accepting the future trajectory we are now barreling toward. To create that future, we must author and cocreate new stories that help direct it.

Envisioning the Future

In a photo, seawater laps at the front steps of the house. In an aerial animation, seawater creeps into Manhattan under a high-emissions scenario. A map peers sixty years into the future to show that a high-emissions scenario makes life in North Carolina more like Florida, and Minneapolis feels more like Kansas. An increasing array of software and graphics helps you visualize your region's future and makes climate impacts part of your social reality, as do scores of urban and land use planning documents that identify weak links—and strengths—in our institutions, systems, and communities.

Seawater at the front steps is one vision of the future, but by itself it can be disengaging and even paralyzing. UK sustainability strategist Jem Bendell has a useful exercise to envision choices needed to lead us to our desired future.[58] First of all, how do we keep what we really most want to keep? If a collective list includes almond trees, laptops, and clean air, *how* do we keep those? The chains that create and sustain those objects are long and involve tough choices about land, earth elements, values, humans, and Others. If we want to keep laptops, how could they be built to last and be repairable and update-able, and also not mistreat those who make them for us? Second, what do we need to let go of (collectively) in order not to make matters worse? Is it gas-powered cars and fossil fuel subsidies, or meat at every meal? Third, we need to restore the attitudes and approaches to life and organization that fossil fuel culture has eroded. That could include shifting our identities beyond the things we own; building connected, supportive communities; changing our expectations for instant gratification; and revaluing the lands and atmosphere that sustain us. Bendell's final step is reconciliation—with past mistakes, with each other, and even with a higher power. Without this deep, inner adaptation and acceptance, he says we risk tearing societies apart.

A Greek proverb says, "A society grows great when old men plant trees whose shade they know they shall never sit in." The proverb's lesson is about forward thinking and being selfless and generous, dreaming into the goodness of a fully grown tree. A similar metaphor is "cathedral thinking." Cathedrals and other ancient structures were built over many generations, and the individuals who began work on them often didn't live long enough to enter the doors. But that didn't stop people from drawing up plans, laying foundations, forming walls and spires, and painting the ceilings. Cathedrals were for the future. The climate crisis requires similar generous future-thinking. It shows that we care that humanity continues to flourish, long after we are gone. It shows that we want to be remembered as better ancestors.

There are hundreds of examples right now of forward-thinking, and some are mentioned in this book. The carbon-neutral pledges signed by many cities and countries are also plans for the future—plans that involve deciding how to get to where we want, what we need to let go of, and what shifted attitudes and practices will help us get there.

A museum exhibit in Sweden takes an unusual direction to envision the future: the exhibit looks back from the year 2045, describing the events that led us to that year.[59] The multidisciplinary "climaginaries" who designed the Carbon Ruins exhibit imagined how we would look back from 2045 and view current fossil fuel culture. The exhibit displays cultural "artifacts" such as nylon stockings, plastic Legos, samples of coal and crude oil, "frequent flier" cards, extinct species, and stainless steel produced in 2045 *only* through recycling. It's a fascinating way to picture a future by evaluating objects and practices of the present and recent past. What might agriculture and transportation look like in 2045? What will we eat and wear? What would we choose to separate from and museum-ize? What needs to be enhanced and celebrated?

Authoring New Stories

In her story, Natalia wrote rationally about how climate change came to be. She recognized that humans had shortcomings but also choices and lessons to learn. Her story clearly represented the liminal transition, but it wasn't "her" story, for she was not present in it.

If you perceive your life path as already fully paved by the threat of climate change, it might seem as though your actions, your career, and your identity are preset and determined. Most of us perhaps do not believe we are authors of the climate story or even characters in it who possess agency. Yet claiming authorship in emotionally engaging stories is an important way to actively *live in* the story, to shift how you perceive the world, and narrate it toward brighter trajectories.[60] The creativity of stories (both composing them and listening to them) is an essential way to explore and develop your agency, to engage, and take meaningful action. Here are several ways to claim and embrace your authorship of a climate change story.[61]

First, thoroughly integrate your emotions into your stories, whether about the present or the imagined future. Write about emotional barriers faced and overcome, whether fear, powerlessness, or feeling overwhelmed. As we've noted, emotions are now largely absent (as is moral responsibility) in discourses over climate policy and action. Working through and expressing emotions in stories helps you progress through the rite of passage's difficult transition. Emotions like self-doubt, guilt, and shame hamper a sense of authorship; as discussed, drop the self-directed blame and shame and instead focus emotional energy on "storying" systems and practices that block reincorporation. Emotions of anger and hope can be extremely powerful catalysts to tell stories and take action. As mentioned before, talking about "negative" emotions is more effective in influencing others than neglecting to mention any emotions at all.

Second, use stories to tease apart the complexity of this looming super-wicked issue – rather than allowing its complexity to hinder your authorship. When you think of the climate crisis as a global whole, it's far too big of a story to tell and too buried in scientific rationality. Tell stories that render small pieces of the crisis as personal and relevant, even if fictional—of a wise grandmother who guides the townspeople to new ways of living, of an industry that transforms from a climate-villain to a climate-hero. When writing about the future, you're presenting a long-term vision that identifies a needed change or the gap you think needs to be closed. This kind of creativity is crucial for exploring and developing a sense of agency[62] (a bit like imagining an action before actually undertaking it). An active agent (or author) is not passively influenced by surroundings and old stories, but instead takes a proactive role in shaping new stories and developing reflective awareness. And, stories can help people *believe* they can change things.[63]

And third, tell stories of co-creation and cooperation. The story of climate change needs a whole cast of characters, not a sole hero and a single "bad guy." The story about a world that addresses the climate crisis is a story of millions of collective actors coauthoring and cocreating the future. They cooperate not only with each other but also with the living world as they implement interbeing and interdependency into everyday life. Knowing that you coauthor the future with important social others helps conquer feelings of powerlessness.

If the characteristics of the "old story" of climate change include separation from and mastery of nature, scarcity, domination, growth, and struggle, what are characteristics of the "new story"? Interbeing, harmony with nature, collective co-creation, abundance, postgrowth, and an ease of being.[64] That's a great cast of characters to write and dream about.

Try writing some new climate change stories yourself with the story prompts in box 8.1. Gather important social others around a fire or a candle (with a wick, not a battery) and see what new stories you can tell.

In the same way that "talking works," in the experience of one student, telling new stories works. Andy Lambert, a master's student in atmospheric science, said that until he wrote the first two "new stories" in box 8.1, he saw the future no differently from what exists now. In the old story, he said we were the affected observers. Now, he sees a very different future:

> I am so much more hopeful. The new story is ecocentric. In it, we are actors. We affect things. We change things. The health of the environment around us is as important as the health of the economy and our bank accounts. The new story accepts our past and uses it to propel us into a cleaner future. It depicts the world as something that can be changed by a mass of hopefuls, rather than by a handful of elites. . . . These new stories provide hope for those who feel helpless about climate change. I think most importantly, these new stories provide direction.

Re-storying the climate crisis is a fruitful way to prepare for what might spring up during the transition and dream into the end-place where we want to arrive.

In 2020 as humans spread COVID-19 around the globe, I heard various commentators and friends wonder what we would learn from *this* rite of passage, how we (and our institutions and economy) would be changed. The prognostications varied widely, evidence that divergent paths (both fruitful

Box 8.1 Prompts for Writing New Climate Change Stories

1. Write a "I Have a Dream" speech of climate change. (If you're not familiar with Martin Luther King Jr.'s original, listen to it here: https://kinginstitute.stanford.edu/king-papers/documents/i-have-dream-address-delivered-march-washington-jobs-and-freedom. It is truly an example of a story that makes people believe they can change things.) This is your first-person, emotion-filled, passionate vision and dream for humanity's future. Thoroughly envision your co-creators who help you arrive at this "promised land." To help form your dream, ponder the envisioning steps above.
2. Write a letter to yourself from 2040. Describe for your current-self what life is like in 2040, how you feel, and how we addressed climate change. Describe what artifacts or practices that you and others jettisoned or changed. Describe how you and society decided what was important, what to let go of, and how attitudes shifted. Describe how you became engaged in meaningful ways, the skills you acquired to do that, and advice for your current-self.
3. Choose a potential "danger" that could derail our liminal transition to a post-fossil fuel reincorporation. The danger could be posed by an industry, an individual or group, a technical problem, a habitual practice, or a specific condition in the living world. Tell a story of how your group of protagonists (of which you are a member) sets out to address and then eliminate this danger. Describe what you encounter on your way.
4. Create a story of your climate change vision through pictures and captions. If you're a visual and/or artistic individual, draw or paste up a "vision collage" of your happy, future self (Adapted from https://grist.org/article/the-most-powerful-tool-for-fighting-climate-change/). Ask what kind of future world you want to live in X years from now. Find old magazines (ask a bookstore or variety store when they recycle old ones) and get out the scissors and glue. How will you travel, where will your food come from, what's your job, and what will you do for fun?

and dangerous) are available during any liminal transition. Though the climate crisis and the COVID-19 pandemic seem to run at very different speeds, they are nested stories that share a common root and foundational message: our health and the health of the living world, our Eairth, are one in the same. These are new stories that must be told with haste.

NOTES

1. Heidi Hendersson and Christine Wamsler, "New Stories for a More Conscious, Sustainable Society: Claiming Authorship of the Climate Story," *Climatic Change*, November 28, 2019, https://doi.org/10.1007/s10584-019-02599-z.
2. Gary Paul Nabhan and Mark Klett, *Desert Legends: Re-Storying the Sonoran Borderlands* (New York, NY: Henry Holt, 1994).
3. Although the words "story" and "narrative" are similar, all stories are a subset or kind of narrative, but not all narratives are effective stories. (Kimiz Dalkir

and Erica Wiseman, "Organizational Storytelling and Knowledge Management: A Survey," *Storytelling, Self, Society* 1, no. 1 (2004): 57–73.

4. Brené Brown, *Rising Strong* (New York, NY: Random House, 2017).

5. Robert S. Emmett and David E. Nye, *The Environmental Humanities: A Critical Introduction* (Cambridge, MA: MIT Press, 2017), http://search.ebscohost.com/login.aspx?direct=true&scope=site&db=nlebk&db=nlabk&AN=1613973.

6. Hendersson and Wamsler, "New Stories."

7. P. Y. Lin, N. S. Grewal, C. Morin, W. D. Johnson, and P. J. Zak, "Oxytocin Increases the Influence of Public Service Advertisements," *PloS One* 8, no. 2 (2013).

8. Lisa Cron, *Wired for Story: The Writer's Guide to Using Brain Science to Hook Readers from the Very First Sentence* (New York, NY: Ten Speed Press, 2013).

9. Brandi S. Morris, Polymeros Chrysochou, Jacob Dalgaard Christensen, Jacob L. Orquin, Jorge Barraza, Paul J. Zak, and Panagiotis Mitkidis, "Stories vs. Facts: Triggering Emotion and Action-Taking on Climate Change," *Climatic Change* 154, no. 1–2 (2019): 19–36.

10. Melanie C. Green and Timothy C. Brock, "The Role of Transportation in the Persuasiveness of Public Narratives," *Journal of Personality and Social Psychology* 79, no. 5 (2000): 701–21.

11. Morris et al., "Stories vs. Facts."

12. Ellen Peters and Paul Slovic, "The Springs of Action: Affective and Analytical Information Processing in Choice," *Personality and Social Psychology Bulletin* 26, no. 12 (2000): 1465–75.

13. Stuart A. Kauffman, *Investigations* (Oxford: Oxford University Press, 2002).

14. Ken Baskin, "Complexity, Stories and Knowing," *E-CO* 7, no. 2 (2005): 32–40, p. 38.

15. Baskin, "Complexity, Stories and Knowing," 35.

16. Stanley N. Salthe, *Development and Evolution: Complexity and Change in Biology* (Cambridge, MA: MIT Press, 1993).

17. Nancy Menning, "Narrating Climate Change as a Rite of Passage," *Climatic Change* 147, no. 1–2 (2018): 343–53.

18. Menning, "Narrating Climate Change as a Rite of Passage," 345.

19. Menning, "Narrating Climate Change as a Rite of Passage."

20. Menning, "Narrating Climate Change as a Rite of Passage," 346.

21. Menning, "Narrating Climate Change as a Rite of Passage," 347.

22. Menning, "Narrating Climate Change as a Rite of Passage," 349.

23. Hendersson and Wamsler, "New Stories."

24. Dale Jamieson, *Reason in a Dark Time: Why the Struggle against Climate Change Failed—and What It Means for Our Future* (New York, NY: Oxford University Press, 2017).

25. Stephen Mark Gardiner, *A Perfect Moral Storm: The Ethical Tragedy of Climate Change* (Oxford: Oxford University Press, 2011).

26. Menning, "Narrating Climate Change as a Rite of Passage," 346.

27. Paul Kingsnorth, "Life Versus the Machine," *Orion*, 2019.

28. Renee Lertzman, *Environmental Melancholia: Psychoanalytic Dimensions of Engagement* (New York: Routledge, 2015).

29. Heidi Neidl, "How to Talk to People Who Doubt Climate Change," *Rolling Stone*, March 3, 2020, https://www.rollingstone.com/politics/politics-features/katharine-hayhoe-evangelical-christian-climate-scientist-953086/.

30. Miyo McGinn, "2019's Biggest Pop-Culture Trend Was Climate Anxiety," *Grist*, December 27, 2019, https://grist.org/politics/2019s-biggest-pop-culture-trend-was-climate-anxiety/.

31. Rebecca Willis, "Five Things Every Government Needs to Do Right Now to Tackle the Climate Emergency," *The Conversation*, September 18, 2019, http://theconversation.com/five-things-every-government-needs-to-do-right-now-to-tackle-the-climate-emergency-123344.

32. Irene Lorenzoni, Sophie Nicholson-Cole, and Lorraine Whitmarsh, "Barriers Perceived to Engaging with Climate Change among the UK Public and Their Policy Implications," *Global Environmental Change* 17, no. 3–4 (2007): 445–59.

33. Johanna Wolf and Susanne C. Moser, "Individual Understandings, Perceptions, and Engagement with Climate Change: Insights from in-Depth Studies across the World Individual Understandings, Perceptions, and Engagement with Climate Change," *Wiley Interdisciplinary Reviews: Climate Change* 2, no. 4 (2011): 547–69.

34. Lertzman, *Environmental Melancholia*.

35. Robert J. Brulle, "From Environmental Campaigns to Advancing the Public Dialog: Environmental Communication for Civic Engagement," *Environmental Communication: A Journal of Nature and Culture* 4, no. 1 (2010): 82–98.

36. https://climateoutreach.org/citizens-assemblies-can-help-build-social-tipping-point-on-climate-change/.

37. Temujin Doran and Thomas Page, "Olafur Eliasson on What Art Can Do to Fight Climate Change," *CNN Style*, August 29, 2019, https://www.cnn.com/style/article/olafur-eliasson-in-real-life/index.html.

38. Julia B. Corbett and Brett Clark, "The Arts and Humanities in Climate Change Engagement," *Oxford Research Encyclopedia of Climate Science*, 2017, doi:10.1093/acrefore/9780190228620.013.392.

39. "So Hot Right Now: Has Climate Change Created A New Literary Genre?," *Morning Edition Saturday, NPR*, April 20, 2013, http://www.npr.org/2013/04/20/176713022/so-hot-right-now-has-climate-change-created-a-new-literary-genre.

40. *Yale Climate Connections*, "Help Wanted: More Climate Storytellers," April 24, 2019, https://www.yaleclimateconnections.org/2019/04/help-wanted-more-climate-storytellers/.

41. A. Cunsolo Willox, S. L. Harper, J. D. Ford, V. L. Edge, K. Landman, K. Houle, S. Blake, and C. Wolfrey, "Climate Change and Mental Health: An Exploratory Case Study from Rigolet, Nunatsiavut, Canada," *Climatic Change* 121, no. 2 (2013): 255–70.

42. Birte Loschenkohl, "Comedy Can Help Us Tackle The Climate Crisis – Here's How," *DeSmog* (blog), December 29, 2019, https://www.desmogblog.com/2019/12/29/comedy-can-help-us-tackle-climate-crisis-here-s-how.

43. "Kids Explain Climate Change to Donald Trump," *Jimmy Kimmel Live*, January 30, 2019, https://www.youtube.com/watch?time_continue=6&v=0sdqNR_s6LQ.

44. P. R. Brewer and J. McKnight, "Climate as Comedy: The Effects of Satirical Television News on Climate Change Perceptions," *Science Communication* 37, no. 5 (2015): 635–57.

45. Chris Skurka, Jeff Niederdeppe, and Robin Nabi, "Kimmel on Climate: Disentangling the Emotional Ingredients of a Satirical Monologue," *Science Communication* 41, no. 4 (August 1, 2019): 394–421, doi:10.1177/1075547019853837.

46. http://news.trust.org/item/20191025050238-mhs28/.

47. A good summary of recent US theater productions is at www.howlround.com.

48. Robin George Andrews, "Melting Chunks of Greenland Are Bringing Londoners Face-to-Face With Climate Change," *Earther*, December 21, 2018, https://earther.gizmodo.com/melting-chunks-of-greenland-are-bringing-londoners-face-1831258353.

49. Gary Braasch, "Climate Change: Is Seeing Believing?," *Bulletin of the Atomic Scientists* 69, no. 6 (2013): 33–41.

50. Saffron J. O'Neill, "Image Matters: Climate Change Imagery in US, UK and Australian Newspapers," *Geoforum* 49 (2013): 10–19.

51. Saffron J. O'Neill and Sonia Graham, "(En)Visioning Place-Based Adaptation to Sea-Level Rise," *Geography and Environment* 3, no. 2 (2016).

52. See http://nalininadkarni.com.

53. Patrick Devine-Wright, "Think Global, Act Local? The Relevance of Place Attachments and Place Identities in a Climate Changed World," *Global Environmental Change* 23, no. 1 (2013): 61–69.

54. K. P. Whyte, J. P. Brewer, and J. T. Johnson, "Weaving Indigenous Science, Protocols and Sustainability Science," *Sustainability Science* 11, no. 1 (2016): 25–32.

55. He coined the slogan to promote *Heroes*, the second installment of his Berlin album trilogy, according to this post – https://blogs.bath.ac.uk/iprblog/2016/01/14/tomorrow-belongs-to-those-who-can-hear-it-coming/.

56. Sam Knights, "Introduction: The Story so Far," in *This Is Not a Drill: An Extinction Rebellion Handbook* (London: Penguin Books, 2019), 9–13, p. 13.

57. Ed Miliband, "Enough of the Climate Nightmare. It's Time to Paint the Dream," *The Guardian*, July 4, 2019, https://www.theguardian.com/commentisfree/2019/jul/04/enough-climate-nightmare-paint-dream-inequality.

58. Jem Bendell, "Doom and Bloom: Adapting to Collapse," in *This Is Not A Drill: An Extinction Rebellion Handbook* (London: Penguin Books, 2019), 73–77, p. 76.

59. https://www.climaginaries.org/exhibition.

60. Hendersson and Wamsler, "New Stories."

61. Hendersson and Wamsler, "New Stories."

62. J. Brockmeier, "Reaching for Meaning: Human Agency and the Narrative Imagination," *Theory and Psychology* 19, no. 2 (2009): 213–34; Katherine K. Chen, "Charismatizing the Routine: Storytelling for Meaning and Agency in the Burning Man Organization," *Qualitative Sociology* 35, no. 3 (2012): 311–34.

63. Tom Van Laer, "Climate Change Is Good|Opinion," *Newsweek*, November 22, 2019, https://www.newsweek.com/climate-change-good-opinion-1473296. (Emphasis mine).

64. Hendersson and Wamsler, "New Stories."

Index

Abram, David, 1–2, 99
agonism, 116–18
agonistic dialogue, 116–17
AIDS, 87, 89
air, 2
albedo effect, 9
Albrecht, Glen, 99
Alexa, 25
ambivalence, 86–87
American Psychological Association (APA), 19, 80
Anthropocene, 99
anthropocentric, 36, 99, 172–73, 175–76, 183–84, 186, 194
anthropocentrism, 172–73, 184
APA. *See* American Psychological Association (APA)
apathy, 85
Art and Energy: How Culture Changes, 31

Baskin, Ken, 194
Becker, Ernest, 100–101
behavior change: holistic, 67–68; individuals, 54–65; social norm and, 62–63; voluntary personal, 55
beliefs, 59–60, 62, 64–65, 139–40, 144, 175–76, 193; religious, 149–59
Bennett, Jane, 179

biophilia, 177
bisphenol-A (BPA), 170
black carbon, 9–10
"Blue Marble" photo, 2
Bohm, David, 116
Bowling Alone: The Collapse and Revival of American Community, 73
Brulle, Robert, 36–37, 40, 201
Buddhism, 151
Buddhists, 151
Buell, Frederick, 30, 32
Butler, Judith, 91

Cacciatore, Joanne, 92
Callison, Candis, 63, 156, 159
Camovitz, Daniel, 180
Carbon Conversations, UK, 131
Carey, James, 40
Carter, Jimmy, 160
Center for Environmental Law, 169
CFCs (chlorofluorocarbons), 15
Chenoweth, Erica, 43
Christianity, 149–50, 155
citizen assemblies, 44, 200, 201
Citizens Climate Lobby, 130
Civilian Conservation Corp, 101
Civil Rights movement, 43
Clarke, Jaime, 123
cleaner energy, 19–20

Climate: A New Story, 178
Climate Action Network, 35
climate change, 35–40; carbon emissions and, 6–8, 14; collective action problem of, 57–58; communication, 40–46, 16–20, 26–27; conversations about, 109–31; emotions of, 79–112; evidence, consilience of, 5–7; feedback loops, 9–10; humans and, 7–8; impacts, 9, 19; mental health tolls of, 80–82; moral imperative of, 137–60; movement, 40–46; problem, 14–16; science of, 2–10; scientists and, 8–10; social reality of, 71; solution to, 14–15; world, 10–13. *See also specific changes*
Climate Change: An Evangelical Call to Action, 156
climate communication, new directions for, 16–20
climate communication with arts and humanities, 201–4
climate conversations, 109–31; agreements for, 127; aims, 115–18; challenges, 124–28; collaboration, dialogue as, 116; exercises, 129; listening/building trust in, 119–20; power differences, 116–18; practicing, 128; preparing for, 118–19; problems, 112–15; resources, 128–31; scaling up, 121–23; values in, 123–24; works, 110–12
climate crisis, 18; cultural silence about, 19; emotions and, 18–19; engagement and action on, 20; individuals in fighting, 18; social actors in, 18. *See also* climate change
climate denial, 8–9, 125; movement, 9, 159
climate disasters, 144
climate gentrification, 145
climate justice, 137, 142–45; and climate disasters, 144; and climate gentrification, 145; and climate migrants, 145; intergenerational justice, 147; procedural justice, 147; restorative justice, 147
Climate Justice Action, 147
Climate Justice Alliance, 146–47
Climate Justice Coalition, 147
climate migrants, 145
Climate Outreach, 131, 148–52
Climate Psychiatry Alliance, 81
Climate Reality Project, 130
climate silence, 79, 109, 113
climate *versus* weather, 5
clouds, 2
CO_2 (carbon dioxide) emissions, 6–8, 14, 35, 55
cognitive dissonance, 84
communication, 16–20, 26–27, 40–46, 59–60
confirmation bias, 64, 126
consumerism, 32
Corner, Adam, 123
Cornwall Alliance, 157
Cornwall Declaration, 157
COVID-19, 17, 24n53, 38–40, 175, 198, 208–9
cultural transformation, 39–40, 46–47
cultural trauma, 37–38
cultures, 26
Cunsolo, Ashlee, 91, 202

Damasio, Antonio, 84
deforestation, 6
denial, 85, 92–97; implicatory, 64–65; socially organized, 64
Desert Legends, 192
dharma, 150
dialogue, 116–18. *See also* climate conversations
Dickinson, Janis, 95–96, 100–101
direct action protests, 41–42
disavowal, 85–86
distal defenses, 96–97
dynamic obsolescence, 33

Eairth, 1–20; atmosphere, 2; climate change, science of, 2–10; climate-changed world, 10–13; climate

communication, new directions for, 16–20; global temperatures, 5–6; greenhouse effect, 4; growth economy, 172–75; indigenous people and, 180–81; interbeing, 178–80; interrelationship, 178–80; problem, 14–16; relationship with, 167–86; Rights of Nature movement, 181–83; switching beliefs, 175–76; switching fuels, 175–76; troposphere, 3–4; weather, 5, 13–14
Eairth Citizenship, 20, 183–84
Earth Guardians, 146
Earth Overshoot Day, 171
eBird project, 11
ecoAmerica, 130
ecocentrism, 176–83
ecological habitus, 37
ecological modernization, 41
Einstein, Albert, 167–68, 186
Eisenstein, Charles, 178
electric slaves, 25
emotions, climate change and, 79–112; ambivalence, 86–87; apathy, 85; changed normal, 98–100; defense mechanisms, 84–87; denial, 85; disavowal, 85–86; ecological grief, 90–92; facing, 82–84; fears, 82–84; fellowship, 102; hero projects, 100–101; mental health tolls, 80–82; mourning, 90–92; resilience, 97–102; system justification theory, 93–94; terror management theory, 94–97; theories of denial, 92–97; working through, 87–89
empathy, 88
Enchanting a Disenchanted World, 32
Encyclical on Climate Change and Inequality, 153–54
Endangered Species Act, 182
energy, 27–32; converters, 28; source, 28–31; system, 27–28
environmental groups, radical *versus* reformist, 41–42
environmental justice, 142–45
environmental melancholia, 90

Environmental Melancholia, 80
environmental racism, 143
environmental social movement, 40–42
Essential Partners, 131
ethics, 138
Evangelicals, 53, 150, 156–57, 159
Evans, Coral, 145
Extinction Rebellion (XR), 43–44
ExxonMobil, 9

faith traditions, 149–57; Buddhism, 151; Christianity, 149–50; *Encyclical on Climate Change and Inequality*, 153–54; evangelicalism, environmental spilt within, 156–57; Hinduism, 150–51; Islam, 150; Judaism, 151–52; religion, pivotal role of, 152–53; religious environmentalism and its opposition, 154–56
family group, 71–72
fears, 82–84
Feygina, Irina, 94
Fisher, Dana, 45
Flight Behavior, 97, 100, 102
forest fires, 4, 9–10, 13
fossil fuel culture, 25–47, 194–96; climate change, 35–40; climate change movement, 40–46; communication, 40–46; consumption, 32–35; fossil fuels as rite of passage, 194–96; history, 27–29; identity, 29–32; social order, 35–40; transformation, 46–47; trauma, 35–40; values, 29–32
fossil fuels, 4, 14, 18; coal, 4, 6, 27–31, 82, 127, 169, 206; gas, 4, 6, 8–10, 15, 25, 37, 55, 169, 206; gasoline, 4, 8, 26, 28, 186; greenhouse gases, 4, 6, 13, 15, 25, 35, 53, 55; natural gas, 4, 6, 10, 27–28, 31; oil, 4, 6, 9, 27–32, 37, 55, 169, 206; wood, 6, 28
fuels, 27, 175–76. *See also* fossil fuel culture
Fuller, Buckminster, 42

Gaesser, Ray, 112

Ganesh, Shiv, 116
gas flaring, 55–56
Ghosh, Amitav, 26, 30, 69, 174
glaciers, 5–6
global warming/heating, 4
Global Warming's *Six Americas*, 16
Google Assistant, 25
gratitude, 181
The Great Derangement: Climate Change and the Unthinkable, 26
Great Turning, 184
greenhouse effect, 4
greenhouse gases, 4, 6, 13, 15, 25, 35, 53, 55
Green New Deal, 39, 56, 94
greenwashing, 37
grief, ecological, 90–92
Griffiths, Jay, 98–99

Haidt, Jonathan, 140
Handley, George, 158–59
Haraway, Donna, 99
Hayhoe, Katharine, 20, 83–84, 100, 112, 122, 131, 199
Hinduism, 150–51
Hindus, 150–51
Hine, Thomas, 32–33
Hoffman, Andrew, 46, 126
How Climate Change Comes to Matter: The Communal Life of Facts, 63
Humilocene, 99

ice, 5–6, 9
immortality, 100–101
Indigenous Environmental Network, 147
indigenous people, 152, 180–83
individuals, 53–74; additional factors, 56–58; collective citizens, 69–73; collective problem, 69–73; constraints, 56–58; family group, 71–72; information and, 58–61; interactions to social change, 65–69; outdoor enthusiasts, 72; personal change *versus* social change, 55–56; role of, 54–58; as social actors, 61–65; in social groups, 61–62; social groups dynamics, 63–65; social norms and, 62–63; workplace group, 72–73
information, individuals and, 58–61
information deficit model, 59
interbeing, 178–80
International Climate Justice Network, 147
interrelationship, 178–80
Islam, 150
I Want That!, 32

Judaism, 149, 151
Juliana v. United States, 146
"just transition," 145–49

Kearns, Laurel, 155–56
Kennedy-Williams, Patrick, 79–80
Kimmerer, Robin Wall, 177, 181
Kingsolver, Barbara, 97, 202
Kirkman, Robert, 113
Klein, Naomi, 144
Kübler-Ross, Elisabeth, 91

Lambert, Andy, 208
Leopold, Aldo, 177
Lertzman, Renee, 80, 86–87, 121–22, 198
liminal transition, 197–200
listening, 119–20, 128
litigation, 146
Living Room Conversations, 131
Lord, Barry, 31

Machiavelli, Niccolo, 35–36
Macy, Joanna, 184
Malibu, Rich, 144
Mann, Charles, 27–28
Mann, Michael, 56
Maori tribes, 182
material culture, 30
Menning, Nancy, 194–96
mental health challenges, 80–82

methyl-mercury, 167, 170
mindfulness, 35
Moore, Kathleen, 141
moral: abstractions, 141–42; duty, 140; foundations and arguments, 138–40; imagination, 142; intuitions, 140
moral imperative, climate change, 137–60; climate justice, 142–45; environmental justice, 142–45; faith traditions, 149–57; foundations and arguments, 138–40; frame and imagination, 140–42; "just transition," 145–49; religious conversations, strategies for, 158–59; urgent, 159–60
mortality, 95–96; distal defenses, 96–97; proximal defences, 95–96; psychological defenses, 95–96
mourning, 90–92
Muslims, 150

NAACP (National Association for the Advancement of Colored People), 146
Nabhan, Gary, 192
Neimeyer, Robert, 92
Nelson, Michael, 141
Nordhaus, William, 171
Norgaard, Kari, 36–37, 40, 64, 67

On Dialogue, 116
Oreskes, Naomi, 9, 153
Orwell, David, 173
ozone hole, 15

Pacha Mama, 182
Pachauri, Rajendra, 59
partitioning, 114–15
phenology, 11
plastics, 168–71
Pope Francis, 153–54
premeditated waste, 33
The Prince, 35
Pritchard, Daniel, 127

psychological defenses, mortality and, 95–96
Putnam, Robert, 73

Qur'an, 149, 150

Rahman, Jamal, 149, 160
Randall, Rosemary, 34, 90
rational hope, 20, 122, 196
reciprocity, 181
reincorporation, climate change, 205–6
religious environmentalism, 154–56
Rights of Nature movement, 181–83
Rising: Dispatches from the New American Shore, 69
Ritzer, George, 32
Roberts, David, 102
Roosevelt, Franklin, 101
Rose, Jonathan, 173
Rush, Elizabeth, 69

Safina, Carl, 100
sand timers, 128
Santana, Carlos, 99
Schmidt, Laura, 92
Schori, Katharine Jefferts, 158
Schwartz, Shalom, 123
scientists, 8–10
Scott, Rick, 17
The Secret to Talking About Climate Change, 120
self-efficacy, 59, 70, 102
self-esteem, 96–98, 100–102
Simard, Suzanne, 180
Singer, Peter, 140
Siri, 25
smart assistants, 25
smart devices, 25
social actors, individuals as, 18, 53–74; additional factors, 56–58; collective citizens, 69–73; collective problem, 69–73; constraints, 56–58; family group, 71–72; information and, 58–61; interactions to social change, 65–69; outdoor enthusiasts, 72; personal

change *versus* social change, 55–56; role of, 54–58; as social actors, 61–65; in social groups, 61–62; social groups dynamics, 63–65; social norms and, 62–63; workplace group, 72–73
social change, 35–40, 43, 45, 65–69
social groups, 61–62; dynamics, 63–65; individuals in, 61–62
social inertia, 36–37
social movement, 40–46
social norms, 57, 62–63; descriptive, 62–63; injunctive, 62–63
social order, 36–37
social reality, 70–71, 115, 118–19, 205
solar energy, 3, 27
Soleil, Natalia, 191–92
Speth, Gus, 160
spiral of silence, 113–14
Stone, Christopher, 181–82
stories, climate change, 191–209; engagements, 200–205; fossil fuels as rite of passage, 194–96; liminal transition, 197–200; new stories, 207–10; old stories, 196–97; in our lives, 192–94; reincorporation, 205–9
stress, 82
symbiocene, 99
symbiosis, 99
symbolic culture, 30
system justification theory, 93–94
system justifiers, 94

talking stick, 128
terror management theory, 94–97
3.5 percent rule, 43
Thunberg, Greta, 45, 137
Tostevin, Bob, 173
Transition Town Movement, 39
TRAX system, 94
troposphere, 3–4

ultraconvenience, 33
USA National Phenology Association, 11

Vibrant Matter, 179
virtue ethics, 140
virtue signaling and shaming, 56

weather, 5, 13–14
wildfires, 4, 9–10, 12–13, 17, 71, 79–81, 84, 86–87, 90, 92, 94, 102, 114, 144–46
Williams, Rowan, 176, 185
Wilson, E. O., 177
wood wide web, 180
Worden, William, 92
workplace group, 72–73

Yosemite, Ansel Adams (photo), 2
Youth Climate Movement, 45–46

zika virus, 114–15
Zoller, Heather, 116

About the Author

Julia B. Corbett is a professor in the Department of Communication and Environmental Humanities Graduate Program, University of Utah. Before receiving her PhD at the University of Minnesota, she was a reporter, park naturalist, and natural resources information officer. She summers in her mountain cabin in western Wyoming. Corbett writes about human relationships with the living world, exploring the influence of human culture on our deep interdependencies with Others. Her books include *Out of the Woods: Seeing Nature in the Everyday* (winner, Reading the West Book Award in Nonfiction), *Seven Summers: A Naturalist Homesteads in the Modern West*, and *Communicating Nature: How We Create and Understand Environmental Messages*.

www.ingramcontent.com/pod-product-compliance
Lightning Source LLC
Chambersburg PA
CBHW070829300426
44111CB00014B/2495